MW00620604

Endorsements

Why is the problem of toxic leadership so intractable and difficult to diagnose? Because the concepts of management and leadership are so often confused. Schultz and Williams sort out this crucial difference and emphasize the importance of leadership in creating a long-term productive organizational culture.

Thomas Anthony
Director, Aviation Safety and Security Program
Viterbi School of Engineering
University of Southern California

Jim and Chuck have provided in this book an important contribution to business leadership literature. They ask the reader to go beyond determining business success through examination of balance sheets and income statements alone, by highlighting the importance of identifying and repairing group dysfunction. Their history of success in achieving positive change in large organizations by making safety a core value is clearly evident in their writing. The culture of all businesses change as each new generation enters the workforce. The principles outlined in the book will help the manager/reader understand how to adapt and maximize the organizations' effectiveness in very practical ways. This book is not about theories or philosophies of leadership; it's a practical guide on how to identify the needs of an organization and the fixes necessary to make it highly successful by two people who have actually done it.

J. E. 'Ed' Codd
Operating and risk management executive, retired
Ford Motor Company, CSX Corporation, Waste Management Inc.

Jim Schultz and Chuck Williams have given us a practical guide to driving change, focused on the for-profit corporation—a focus that spotlights the *tough* in tough love. But the book suggests lessons that may be adapted by anyone seeking to lead an ailing organization out of its misery and into a more useful future. Veterans of organizational wars will recognize the challenges they address, and beginners who read closely may be able to skip some of the mistakes most of us made as we sought to gain traction. The authors' strong message about safety as a core value, operationalized with the participation of the entire workforce, provides a solid foundation for ethical leadership that should be required reading for all of those seeking to claim the title of 'executive.'

Grady C. Cothen, Jr.
Transportation safety consultant, attorney,
retired federal executive

A book worth reading - many times over! No matter how well we think our businesses are performing, let's not read our own headlines. Every business has some degree of dysfunction. A leader's role is to root it out. Can you imagine a profitable and safe company prospering amid dysfunction? It won't happen. Jim and Chuck provide the perfect roadmap for a company of any size to perform at an optimal level. For example, their roadmap to overcome dysfunction and turn the culture is at the same time simple, profound and actionable. Any leader can accomplish this if they have the grit and desire. The authors are first and foremost world class practitioners. I've personally witnessed how they helped a Fortune 500 company overcome massive dysfunction and begin a journey to excellence. This book has become my "go to" manual to which I will refer often.

Brian Fielkow
Chief Executive Officer
Jetco Delivery

Every organization possesses some degree of dysfunction. To think otherwise is foolhardy and a denial of the truth. In the pages that follow, Jim Schultz and Chuck Williams have meticulously described organizational dysfunction in simple and understandable terms. The authors clearly define a roadmap to success based on factual analysis of the causes of the dysfunction and offer proven methods to affect and sustain a turnaround. Fixing a bad company is a daunting task. It takes a wise and committed leader to change the values, culture and attitudes of the people. If your organization is stuck in the quagmire of mediocrity and paralyzed by a resisting workforce, this book if for you. If you have the guts to say *stop* and the willpower to make a change, this book is for you.

Al Gorthy
Captain, United States Navy, retired

In this book Jim Schultz and Chuck Williams address the most critical issue facing the today's new leaders: how to recognize a dysfunctional organizational culture and what to do about it. As someone involved with the turnaround of a number of organizations, I know that the often-unspoken truth is that things are usually worse than anyone in the organization will admit or even realize. Jim and Chuck have clearly laid out the leadership path - from a culture of failure to a culture of success – in a way that the reader can use in almost any situation. This guide is essential reading for anyone interested in leading others irrespective of organizational role.

George Avery Grimes, PE, PhD
Transportation Research Board Rail Group Chair

Leaders and managers of every company or organization, regardless of mission, size or charter, will benefit from this book by Jim Schultz and Chuck Williams. What type of leader are you and what type of leader do you aspire to be? Peering through the mirror of this book, you will see a reflection of yourself and your organization for what it is, and, more importantly, what it can become. Leadership is not a solitary endeavor; it is a team-focused undertaking that requires engagement, nurturing, vision, respect and perseverance. Great leaders recognize the distinct difference between power and authority. While power may be implied by title or position, it alone is largely impotent. Authority, on the other hand, is the power earned through respect, truthfulness, transparency, accountability, and the maintenance of a just culture. A strong and effective leader is one who has earned the respect of those being led. That is the path to overcoming organizational dysfunction and building a sustainable, high performance enterprise. This book provides a roadmap to that end and is a must-read for leaders and managers of every organization from Fortune 500 corporations, to small businesses and nonprofits, and within government.

Rick Inclima
Director of Safety and Education, retired
Brotherhood of Maintenance of Way Employees Division
International Brotherhood of Teamsters

A to the point guide to maximize total organizational performance. In these pages, Jim and Chuck provide real life strategies and tactics that will produce a vibrant and productive culture that benefits all stakeholders. It is a must read for every manager.

Sharon Joseph
Founder and Principal
James McAfee
Registered Representative
Joseph Financial Partners

Chuck and Jim use their many years of turnaround experience to cut through the shelves of leadership and so-called business how-to books. They provide a brutally honest way of telling when an organization is in trouble and how to identify the traits of its bad actors. When you read this book the faces in the organization, both good and bad, will appear with each paragraph. Excellent work gentlemen, thank you for your candid insight!

Dennis D. Marzec
President
Patriot Rail Company LLC

After teaching and conducting safety and reliability research for more than 30+ years on complex, large-scale technological systems, I believe that safety culture is analogous to human body's immune system that protects against pathogens and diseases. As the renowned professor (emeritus) of Psychology at the University of Manchester, James Reason who is considered as a founding father of understanding of root-causes of human error and systems failures has succinctly said, because of the pervasive nature of safety culture and its widespread impact "it can affect all elements in a system for good or ill." Schultz and Williams have skillfully utilized their impressive professional experience, as organizational "doctors" and "pathologists", and meticulously addressed pathogens and diseases that can afflict a company and how it can be prevented for cured. This illuminating book, specially its analysis of "culture of failure", is *highly* recommended and essential reading for immunizing, protecting, maintaining, and recovering the healthy culture of any company.

Najmedin Meshkati, PhD, CPE
Fellow, Project on Managing the Atom, Belfer Center for Science and International Affairs, Kennedy School of Government – Harvard University
Professor of Civil/Environmental Engineering; Industrial and Systems Engineering; and
International Relations – University of Southern California
Commissionaire – The Joint Commission

Chuck Williams and Jim Schultz have written an eloquent guide for those who cherish value driven leadership, integrity and organizational excellence, based on their experiences leading organizations over lengthy and successful careers. They thoughtfully discuss identifying and strengthening dysfunctional organizations, not merely helping good companies become great ones. Their wisdom and advice are appropriate across many disciplines. As one example, leadership in healthcare might glean new insight and inspiration about how to strengthen organizational culture and improve safety, patient care, and provider satisfaction while curtailing healthcare system dysfunction.

John W. Overton, Jr. M.D
Surgeon and pilot

Jim and Chuck give you the playbook on how take your company back from the 'rebels & terrorists' and transform a dysfunctional poor performing organization into a winner!

Brad Pinchuk
President & CEO
Hirschbach Motor Lines

When your organization is dysfunctional and in chaos you need sound and logical techniques to turn it around. In this book, Chuck and Jim provide a guide to make it happen based upon their decades of high-level experience transforming poor cultures.

Ab Rees
Chairman and CEO, retired
Railworks Corp

How many will look at this book and say to themselves that it's just another book covering topics already covered? For the sake of your company and your career, I challenge you to read and absorb the real-life lessons offered here. This work is a recipe for success that all managers should follow. The success of any company begins with establishing an excellent safety culture and maintaining it. As so correctly pointed out this book helps managers recognize dysfunctional creep before it can take hold and disrupt institutional 'blood flow.' Dysfunction can manifest itself in many ways. I applaud this book and know that it will take its place as one of those 'must read' lessons for future generations.

Timothy L. Smith
Chairman Emeritus
National Association of State Legislative Board Chairmen
California State Legislative Board
Brotherhood of Locomotive Engineers and Trainmen

Jim Schultz and Chuck Williams have created an outline for reviving failing operations. Making safety the cornerstone value in the decision-making process strongly reflects appreciation for the company's greatest asset, its people. Instilling a just culture into the operation brings accountability and success, and this book simplifies that project.

James A. Stem, Jr.
National Legislative Director – Transportation Division, retired
International Association of Sheet Metal, Air, Rail, and
Transport Workers (SMART)

Williams and Schultz have pulled together the perfect follow-on companion book to Larry Bossidy's and Ram Charan's book *Execution*. They provide an eye-opening and honest approach to addressing the human factor dysfunction that can derail the best corporate strategy. These well-articulated people strategies are sharp reminders of common problems managers tend to want to overlook and hope solve themselves over time. Thank you for sharing your wisdom from years of hands-on experience to help others become more effective leaders.

Jennifer G. Whiteman
Chief Financial Officer
Patriot Rail Company LLC

Bad Company/Good Company
A Leader's Guide:
Transforming Dysfunctional Culture

BAD COMPANY/ GOOD. COMPANY

A Leader's Guide: Transforming Dysfunctional Culture

CHARLES E. WILLIAMS AND JAMES T. SCHULTZ

MCP BOOKS

MCP Books
2301 Lucien Way #415
Maitland, FL 32751
407.339.4217
www.millcitypress.net

Printed in the United States of America.

LCCN: 2019910093

ISBN-13: 978-1-54566-145-1

Dedication

To the people in our business lives who influenced and helped shape our approach to good governance, leadership, and how to effectively guide organizational change.

Our special thanks to Lisa and Constance for their unceasing support and encouragement in life's journey, with special acknowledgment of their advocacy and endurance through many long hours of sequestration while we were writing and collaborating on this manuscript.

Chuck Williams
Jim Schultz

Table of Contents

Foreword

I t was a bad sign. It was my first day on the job as the new CEO of one of the nation's busiest and most complex transportation entities, coming off a catastrophic safety event that had resulted in fatalities and injuries. I could sense that the culture was one of defeat and resignation. Then, literally hours after I sat down at my desk on that first day, a group of external auditors was at my door, and they had a lot to say. They presented multiple pages of deficiencies and longstanding gaps in governance, and they wanted to know what I was going to do about it. I asked them to give me a few days to digest their report and do a quick assessment of the situation, but it didn't take me long to realize that the auditors were right. And, as I quickly determined, the deficiencies that the auditors documented were mere symptoms of an underlying culture of failure.

Early on, senior staff told me that everything was fine—that the auditors were wrong. Any problems, I was told, were due to outside influences that didn't understand our business. In other words, they gave themselves a pass on accountability—we were, they said, "victims" of forces beyond our control.

It was clear that my staff suffered what Jim and Chuck call the "myth of good." Despite the facts, despite the obvious underperformance and public condemnation, senior staff thought that they were good. Given the complexity and depth of the dysfunction and the pressing need to turn performance quickly, I knew that I needed help. I called on Jim and Chuck to help me assess everything and begin the long, difficult journey to transform a highly reluctant group that believed there was no need to change. They were content to fail and blame others for the outcomes.

Working together, in just a little more than twelve months, using many of the strategies and tactics Jim and Chuck describe in this book, the organization totally turned performance around and embraced a winning culture. So, what did we do? We eliminated toxic people, we instilled a just culture, and we went to an engaged and aligned leadership model with accountability. We focused on day-to-day execution of the basics supported by simple metrics, we established guiding coalitions to tackle thorny issues, and we involved all stakeholders, especially front-line workers and managers in the process, and we listened to them. The results: dramatic improvement in financial, productivity, and operating performance, along with elevated favorable public optics around our presence in the community. Customers responded with high engagement and we grew the business while cutting extraneous costs. It was a magical transformation in record time that was praised in the press, by elected officials, and by federal and state regulators.

Jim Schultz and Chuck Williams are the real deal. I originally met Jim and Chuck in 2001 and saw firsthand how they effectively moved the culture of a complex, resistant, and disinterested company to one of high performance and accountability. I was privileged to join them as they worked to instill a culture of safety and success at a Fortune 200 international transportation and service company that not only had appalling safety performance but also produced disappointing results in most other areas of business execution. Here, we saw in action the damage corporate incompetence and managerial malaise can inflict. We validated our belief that poor safety is a leading indicator of an overall failed culture that is characterized by avoidance of reality, cronyism, lack of meaningful performance metrics, open resistance to change, chronic short-term focus, inconsistent values, ambivalent strategy, and a mind-set of deflection instead of reflection, which led to a wholesale lack of respect for employee contributions. Using Jim and Chuck's approach outlined in this text, we addressed the technical and human aspects of the defective processes, we built and initiated strategies that substantially improved revenue return and profits, and at the same time we improved safety performance and overall employee alignment in our field operations. Perhaps the most important takeaway was that leaders shape a company's culture—be it good or bad.

As Jim and Chuck mention early on, "A bad company is a formidable beast. You cannot just bully it into becoming a good company." Such companies, and the people within them, excel at resisting change. Many leaders fail to realize they have a dysfunctional organization until damage has been done. They know some things need improvement and will attempt to fix a few problem areas, but they fail to see the seriousness and true depth of the problems. Early recognition of an organization's state of performance is critical to the success or failure of any leader. It requires a methodical organizational "physical exam," with a baseline review into every department in the company. Every door must be opened and every process inspected to detect the onset of dysfunction.

As a CEO involved in the transformation of several organizations over the past ten-plus years, I know that rebooting a dysfunctional company is not for the faint of heart or the unprepared. Over the years, I have learned that you can't freelance organizational change of this magnitude. At a minimum, a turnaround CEO must have a plan, and must possess a willingness to take on sacred cows, slay them, and set high standards of performance while displaying a positive leadership presence that inspires alignment. Successful leaders are obsessed with the truth, they are value-driven, and they have empathy for those they lead. Doris Kearns Goodwin, in her recent book on leadership, points out that good leaders have "an ambition for self that becomes an ambition for something larger" and, in my view, can influence their team to become part of something larger than themselves.

The turnaround CEO must not only have the traits of a good leader, but he must be extremely well armed for the battle. Jim and Chuck's book provides a solid road map for any leader facing the challenge of fixing a troubled and inefficient company. While theirs is an unvarnished review of the depth of the problems that such a leader will face, it is also a guide, based on years of insight gained and lessons learned, for identifying and applying essential principles for building and sustaining a culture of success.

I wish I had this guide years ago to help me understand what I was facing and what to do about it. Jim and Chuck have encapsulated the fundamental principles of rebuilding any organization that is deeply troubled, or for correcting course for one that is starting to exhibit early onset of dysfunction.

This guide is essential reading for any leader, for anyone who wants to be a leader, and for anyone who seeks to understand and make positive change within their own organizations.

John E Fenton
Chief Executive Officer
Patriot Rail Company LLC
Jacksonville, Florida

Preface

This manuscript is a practical manual offered to help the reader identify companies that are in failure mode, understand how they are failing, and then demonstrate how these bad companies can be transformed. This book has application at all levels of the organization—from the front line to the boardroom.

In its most basic state, we found corporate life to be a cacophony of interactions and relationships, with multiple voices pushing and pulling on organizational mores, traditions, standards, and processes—often leading to dysfunctional onset. We know that organizations do not fail by accident. They fail because of disengaged leadership, ineffective values, lack of disciplined execution, ambivalent priorities, the questionable quality of people that they hire, and the way they manage those people and the work they do.

This is not an academic or theoretical discussion. What follows in these pages is a practical guide to transforming deficient cultures, based upon what has worked for us in our combined ninety-plus years—how we learned to recognize dysfunctional onset, successfully correct course, and anchor a culture of positive energy.

Finally, while the authors are aligned in approach, philosophies, and values, they do have different writing styles. Throughout this book, our opinions and experiences are offered from personal involvement. We decided that the sum of the parts is what is most important. Although each chapter was a fully collaborative effort, rather than try to alter either style and risk loss of personal feel, we identify at the start of each chapter the chapter's primary author.

Introduction

Dysfunction: Abnormal or unhealthy interpersonal
behavior or interaction within a group.

—Merriam-Webster Dictionary

Jim: Imagine that a company had an embezzler skimming money and fouling financial earnings. There is no doubt that there would be immediate and aggressive management intervention. Yet, when managers allow dysfunctional routines to take root in an organization, even deeper financial harm can be done than would be inflicted by an embezzler. While both need to be stopped, dysfunction is a particularly insidious plague because it not only drains the bottom line through unproductive and bungled processes; it also burdens the workforce by complicating job tasks, harming morale, and marginalizing their ability to fulfill company obligations to customers and shareholders.

In this book, we present a road map that has worked for us to provide better corporate governance, higher productivity, and improved margins. A bold claim, to be sure. But as battle-hardened practitioners with ninety-plus years combined tenure circulating through organizational trenches, we have witnessed in living color how dysfunctional routines hinder individual and group performance.

In Sum: Connecting the Dots

In the first three chapters we talk about the ingredients of dysfunction. Starting in chapter 4, and expanding through chapters 5, 6, 7, and 8, we outline why there must be the right leadership, the right team mind-set, and the right culture in place before a company can self-repair. Once there, as we examine in chapter 9, the next step is to design and execute the right processes—the key initiatives that add structure. Following, chapters 10, 11, and 12 go into more detail about central systems that are in place in all good companies. In chapters 13 and 14, and the conclusion, we wrap up, review, and reemphasize some key elements that help management sustain a dynamic culture of engagement and productivity.

A New Generational Model

In writing this book, we were mindful of the fact that change is under way in organizational dynamics. As baseball legend Yogi Berra was credited with saying, "The future ain't what it used to be." Yogi was right. Today, we are in an era where conventional so-called command and control and longstanding hierarchical models are being challenged by new generations of workers and managers who question time-honored approaches to authority. And while still short of quiet panic, we have witnessed hand-wringing in the halls and executive suites in conventional organizations asking: *How do we deal with the millennial, Gen Next, and upcoming Gen Z workers?* The change phenomenon has also become a popular topic, upon which many academics and others like to opine.

While most of our careers were spent in the traditional regimented organizational model, the move to more open and participative governance actually began during the last few years of our time as full-time senior managers in American industry. In addition, over the past ten years in our consulting and coaching work, we have seen firsthand the new generation in action in diverse roles and industries. From what we have seen so far, we are optimistic. We believe that the strategies and observations we offer in the following chapters have application in all genres of governance. In substance, we believe that:

- Great leaders don't sweat the generational encounter. They know that, like any group of new workers, there are managerial challenges and perhaps differing views of work priorities, but they also know that the new generation brings good things to the table, such as high energy, renewed focus on principled organizational conduct, unique and valuable skills, and the desire to go about things the right way.

- Great leaders, those with progressive and engaging leadership traits, already manage with an enlightened touch. For example, they understand and manage with a progressive catechism—i.e., a methodology we call $\underline{V} + \underline{I}^2$ (ensuring that people feel $\underline{v}$alued, $\underline{i}$ncluded, and that they have $\underline{i}$mpact in their work). Without thinking about it, good leaders have already gone a long way to smooth the transition in generational cultures.

We had a conversation with a client in the financial sector, seeking advice on what to do differently to manage millennials. Some in the financial sector, along with many in the industrial segment, have deeply embedded traditional authority roots. Our advice to the client: the basic change-management processes we outline in this book still work, like conducting regular employee listening sessions (people must be heard before they can hear), performing objective gap analysis exercises, creating a compelling case for change, engaging the team through a guiding coalition (of both old and new viewpoints), developing a shared vision, internally branding the process, eliminating toxic people, measuring the right things, and celebrating the *skid marks* (small victories in the new culture). In the end, people are people, and by treating people right, you can't lose.

We have confidence in our approach outlined in this text in any organizational change scenario. It is important to us that it works.

We have children and grandchildren who have parachuted, or one day will parachute, into the workplace with vastly different expectations than we had when we started almost five decades ago.

Treatment of Gender

Another change that has manifested itself during our two careers has been the growing involvement of women in the workplace, especially at the executive level. It has been long overdue. Recognizing that the potential audience for this book is now more balanced between both sexes, we have wrestled with how best to treat the gender issue in writing the material. We looked at using "he/she" and "his/her," but it became very cumbersome, especially when employed three or four times in a paragraph or a long sentence. After much discussion, we opted to simplify things by using a neutral gender "he" and "his" with the understanding that, in all cases, gender is not a consideration and that the comments, observations, and recommendations apply equally to both male and female leaders, managers, and employees.

Context and Tenor

While the principles we discuss in this book will benefit anyone interested in avoiding dysfunctional creep and marginalized performance, our approach here is different from most other leadership books in that we focus on major rebuilds vs. tweaking or adjusting culture to go from good to great. We have successfully utilized the strategies presented in this book to overcome standing dysfunction in very difficult and unwilling environments.

The sum of our experience leads us to the conclusion that highly dysfunctional organizations requiring total cultural rebuilds are common. In the following chapters we sought to paint a picture of what pervasive dysfunction looks like, how it invades and cripples a culture, how it manifests in diminished performance, and how it can be fixed. Even though we discuss in these pages the extreme case, the principles and strategies we offer have direct application in cases when the degree of dysfunction may be less and the extent of needed intervention may not be as apparent. The main message of the book remains the same in any instance: Overcoming organizational dysfunction is difficult, it requires complete dedication, and it is a lot of hard work. People typically fail by underestimating what is required. It is best to go into any transformation prepared for a very fierce battle but one well worth the effort.

This book is also a valuable resource for good managers who strive to remain good. In our experience, being a good performer can present a unique set of risks that may lead to problems. It can perhaps best be summed up by Jim Collins in his book *How the Mighty Fall and Why Some Companies Never Give In*. Here Collins points out that "every institution, no matter how great, is vulnerable to decline. There is no law of nature that the most powerful will inevitably remain at the top. Anyone can fall and most eventually do."

The Journey

We decided to document the techniques and strategies that have worked for us in transforming dysfunctional cultures in both private and public-sector organizations, from entry level to the C-suite, from smaller regional entities to big international corporations. Our bottom-line conclusion: recognizing and healing structural dysfunction is critical to cure an ailing culture and chronic underachievement in safety, customer engagement, financial performance, and operational excellence.

As detailed in subsequent chapters, our experience is that complacency and self-contentment will creep into an organization over time unless there is a management obsession with improvement. This book helps managers recognize dysfunctional creep before it can take hold and disrupt institutional blood flow. We know that most underperforming companies don't realize the degree to which they are bound in dysfunction. To borrow a term from aviation safety, these organizations are replete with what has been called the "normalization of deviance" ethic. Bad practices and negativity are so entrenched that they become the norm, and the definition of good is skewed. Left unchallenged, we have seen firsthand how such dysfunction generates negative cultural inertia, toxic work environments, and high-cost employee and customer issues.

In addition, bad companies disappoint in most significant categories. The consequences can be severe: long-term diminished franchise value, lingering financial underperformance, missed opportunities, and ambivalent social standing. Shareholders, creditors, and institutional investors suffer from a chronic dilution of returns. Dissatisfaction can manifest in reduced confidence and investor desertion. Financial and ethical credibility is everything in a world of tightening credit, demanding institutional covenants, and an unforgiving public when it comes to corporate social virtues.

Success in business is survival of the fittest. History has shown us that sooner or later a culturally fit, well-functioning competitor that knows how to maximize performance will come along and the bad company will go away. And we have learned that to effectively evaluate a company, one must look beyond pure bottom-line numbers. As William Bruce Cameron wrote in his book *Informal Sociology: A Casual Introduction to Sociological Thinking,* "not everything that can be counted counts, and not everything that counts can be counted." This concept was validated in a February 2018 issue paper published by the National Association of Corporate Directors (NACD) emphasizing that nonfinancial metrics are increasing in prevalence and importance. The article notes that 71 percent of executives in a recent poll declared that communication to the marketplace about purpose, values, and vision was of equal or greater importance than financial results alone by painting for stakeholders a more holistic view of organizational performance.

Savvy investors recognize this and not only put their money where returns are elevated; they also invest where they see good management governance, socially responsible and ethical conduct, future growth potential, and healthy values. They know that when appraising a company, one needs to look beyond just profit and loss statements because, in some cases, bad companies may make money in spite of themselves. This can happen for any number of reasons. Perhaps it is a monopoly, or a short-term trend, or it may be because it is the best of the worst in a field of inferior players. Whatever the circumstances, the margins that are produced are almost always subpar versus potential. The stock languishes, capital is constrained, and innovation is wanting. We saw this happen in a dysfunctional company that made money because of sheer size and reach of the franchise where people had few viable alternatives. Even then, despite positive cash revenues, this company regularly disappointed Wall Street because of habitual dissatisfaction and unrealized earnings potential. As is typical in underperforming cultures, senior management convinced themselves that all was well. It was a prime example of "the myth of good" that we address in chapter 2.

SECTION I

MANIFESTATION OF DYSFUNCTION

Chapter 1

Why This Book Is Needed

Chuck: The world is filled with books on company improvement, so why is *this book* necessary? The answer is very simple: we could not find what we needed in those other books. Most of the company-improvement books that are in the marketplace are in what we call the "good to great" category. In fact, one of the books by Jim Collins is titled *Good to Great*. There is nothing generically wrong with these books, and we believe that many of them can be useful for companies that are good now and want to get better. The GE that Jack Welch inherited was not a bad company but was not a great company either. He and his team made it great, and we have read virtually all the books about Jack Welch and by Jack Welch. Unfortunately, the companies we were trying to fix were not great nor were they good; they were bad. And that is a big difference.

It should be pointed out that there is one book that we believe has application to virtually any company regardless of how well it functions. We have used this book in the past and will make several references to it as we walk our way through the various chapters of this book. That book is *Execution* by Larry Bossidy and Ram Charan. It is a no-nonsense book about getting things done in the workplace and is a great reference for all managers and leaders, no matter how well they think they are doing. *Execution* is not focused on the bad company but, when coupled with this book, provides part of the road map for pulling the bad company out of the quagmire and giving it some basic tools. Chapter

12, "Performance Management," will go into more detail about *Execution* and the messages it presents.

Good companies generally have good managers and good employees who have the necessary skills and attitudes. The appropriate values and practices are in place and the company knows how to operate in its core areas. The leadership is capable and there is a good connection with the next levels down. People can see what is wrong when it is identified for them, and they have the ability to accept and make changes. There may be issues with inefficient processes, wasteful practices, a strategy that needs refocusing, or areas with substandard leadership; but the corrections are generally more adjustments than a complete overhaul. The whole company does not have to be changed and the corrections can be isolated. It may involve adding a process or changing a practice. The transformational message is relatively simple and, most importantly, the infrastructure is basically in place to handle the change and make it all happen successfully. In short, a good company can typically address its problems if it is aware of them and has the desire to fix them. Even if the problems are more massive than normal, good companies have the infrastructure of people, leadership, and attitude to stay the course and be successful. Bad companies do not have the ability to do that. They cannot identify their problems or appreciate the gravity of their problems, and certainly do not have the perspective and the horsepower to fix anything about themselves. They are so insecure and dysfunctional that anything more than doing what they do every day is a bridge too far and not worth even contemplating.

The Bad Company

Classically, dysfunctional companies have huge cultural and systemic issues. The basic pieces are not there or, at best, are only partially there. Leadership is weak, overwhelmed, or misdirected. The corporate office and field operations are typically disconnected and perhaps even at war. Processes are ineffective or nonexistent, and accountability is absent. Personnel are generally inadequate, poorly trained, poorly managed, and not motivated to succeed. There is no effective strategy for success. Any plans for improvement are half-baked and poorly supported, and they fail quickly. What exists is basically a culture of failure where management and employees avoid reality and operate within the illusion that everything is fine. People at all levels are in survival mode and their fear of making a mistake or alienating a person of status is more important than accomplishing something. Not only is the bad company drowning in its own mediocrity, it lacks the ability or the will to recognize its problems and do anything about them. The people in the bad company are generally victims of a relatively new term, "Group Think," where everyone agrees that they are doing it right and everything is fine.

Unlike the good company, a dysfunctional company will not be fixed with a few adjustments. The basic infrastructure of leadership, people, organization, and values is in trouble and does not provide the foundation on which a transformation of any kind can be built. There is very little that you can easily salvage in the bad company, especially the attitudes of management and employees. What is needed is a complete overhaul with virtually everything about the company being on the table. The change process is exacerbated by the fact that you cannot just shut the company down and change everything. You have to run what you have and make the changes that you can on an incremental basis. This approach can slow the transformation process way down from months to years and plays havoc with company morale as change-averse people continuously must deal with change. Getting the company convinced that it needs to transform, can transform, and has the ability to stay the course is probably the most difficult part of the process, because these people are not used to serious conversations about performance and success and have been disappointed by company leaders for their entire careers. The trust level in bad companies is never very high. This needed transformation is a lot more difficult than getting a good company to embrace a new process or even a new strategic direction. It is about creating a totally new company with new attitudes, abilities, and practices.

Fixing the Bad Company

Transforming a dysfunctional company is not for amateurs or the faint of heart. The change process has to be very broad, very deep, and very complex. Virtually everything in the company will be touched by the transformation. It needs to start with the leadership and the company culture. Company management and the employee base that follows them are locked into a set of attitudes that preordain failure. Their value system rejects strategic thinking, operation by process, the importance of good people who are well trained, teamwork, and accountability. The merit system does not exist in the bad company. They are into cronyism and avoiding reality. That all must change. What is needed is a complete refocus with new values, new expectations, and a new attitude about the company and its capabilities. The transformation team must take charge and forge a new success culture that will change the way people think and how they operate.

The most important and most complex part of the transformation involves the people at all levels. It would almost seem easier to shut the place down and change everyone, but this is not feasible for many reasons. First, there are existing business relationships that must be maintained, or at least understood, as well as institutional knowledge within the company that must be saved. Second, some of the existing people, especially well down in the organization, have talent and are worth keeping. Even some of the managers may be worth keeping and will flourish in a more functional environment. The key is taking the time to properly evaluate everyone and being able to find the real keepers. And third, a nearly 100 percent turnover over a relatively short time period is just not manageable, given all the other things that have to take place in a transformation. Undoubtedly a significant number of employees will eventually be asked to leave, but it must happen over a greater time span and after each individual has been properly evaluated. The key to making the people piece of the transformation work is finding the leaders/managers most responsible for driving the existing culture and removing them from the company early in the process. This does not totally clean the slate, but it is

6

a good start and it sends the right message about the seriousness of this process. In most successful transformations, the majority of the key personnel are replaced and this is needed to build new attitudes and practices at the top of the organization and to obtain the leadership and managerial talent that is needed. The employee base cannot be expected to change how they think and operate if they do not see big changes above them. The people piece is key, and this is where many of the critical mistakes are usually made. It is vital that this piece of the transformation gets a lot of attention and is done correctly.

There are no shortcuts to any of this. All that is bad must change or go, and all that is needed must be done. If you do 90 percent of the program, you do not get 90 percent of the results. You generally get mediocrity and possibly total failure. Many change leaders lose their nerve partway through and try to reduce the program or allow bad people to stay way too long. This is all a recipe for failure, as the retained bad people use their best efforts to sabotage everything new and preserve everything old. With so many moving parts, the interrelationships are numerous and highly complex. There is a surprise every day. The potential impacts must be clearly understood before activating any particular piece of the program, and the timing of what is done is generally as important as the actionable items themselves. When things go wrong, the fixes must happen fast and be effective. This all requires much time, much energy, and a high level of commitment.

As if this were not hard enough, the one thing that all dysfunctional companies excel at is being able to resist change. Senior and middle managers have made a career out of sabotaging change efforts. They are old hands at duping CEOs, and slow-playing and outwaiting virtually any effort at change. They know how to isolate transformation messages at the top and have all the reasons for failure and the excuses in place way before any change program even begins to start. The prevailing culture in bad companies is all about survival and keeping things the way they are. If success requires change, they are not interested. These people may not be good at running the company, but they are experts at throwing up obstacles. Transformation leaders need to understand all of this and have a plan to neutralize these people right from the beginning or they will lose.

The reader may think that the typical bad company described in this chapter is some aberration that rarely exists. Unfortunately, this is not true. There are many of these bad companies, and they may actually be in the majority. A respected senior colleague once remarked that good leadership is the exception and not the norm, and that follows for the companies they run. We have seen a number of these companies in the private sector, in publicly traded entities, and in government agencies. Companies are good at hiding their failures and making themselves look good to outsiders who are not really paying much attention. As you will see in the next chapter, CEOs in bad companies support a culture where bad is good and denial about what is happening is normal. They spend a lot of their time concealing or rationalizing their failures. As a savvy businessman once said, "The truth is what I can get you to believe." That definitely works for the bad company.

What Happens to Bad Companies?

As we have pointed out, the collective leadership in bad companies is focused on survival, and they are good at it. Many bad companies continue to plod along. Some make a profit but not anywhere near what they could generate if they functioned properly. CEOs come and go and there may be a reorganization every five or ten years, but the core of bad people generally remains or is replaced by people who continue to perpetuate the failure. Many of these companies eventually disappear or are acquired. Dysfunctional government agencies can go on forever. They just become more bloated and ineffective as they blame their failures on a lack of resources.

In a few rare cases, a top-notch CEO gets involved and the bad company is transformed into a good company and perhaps even a great company. In even rarer cases, that CEO builds such a strong infrastructure of leadership that the excellence continues after that CEO departs, and the company may even achieve higher levels of success.

What This Book Does

As you can see, a bad company is a tough place for someone with good intentions to operate. By the same token, when you make real progress in a bad company, there is no better feeling of accomplishment. You know you have done something unique and of real value, and anyone paying attention knows it as well. This book is basically a handbook on how to identify and understand the bad company, and then how to fix it. We describe the culture, the people, and the practices of a typical bad company. We then go into how a new CEO can turn a bad company around and make it into a good company (we leave the "being great" part for the other books that are available). Most importantly, we point out the pitfalls and mistakes that we have seen well-intentioned CEOs make that lead to failure or to much reduced success. Make no mistake: a bad company is a formidable beast. You cannot just bully it into becoming a good company. That has been tried and the failures are impressive. In those cases, the bad guys won. If a CEO wants to win this war, he must be smarter, tougher, more vigilant, many steps ahead of the opposition, and more committed. Anything less is a waste of time. If you have what it takes, then this book is for you.

Managers vs. Leaders

This topic gets a lot of coverage in the business books, but people are still confused about it. We have all heard that leaders develop the ideas and the managers turn them into reality, or that leaders create strategy and managers take care of the tactics that follow. Jeff Weiner, CEO of LinkedIn, said in an interview with Gayle King in 2017 that managers tell people what to do, and leaders inspire them to do it. In good companies, good leaders are those unique people that make companies special and get them to realize their potential. You pretty much know them when you see them. It follows that leaders are stymied if they do not have good managers who can keep things organized and get things done. In good companies, you have to have both. In fact, great leaders spend much of their time turning their managers into effective leaders today and even better leaders tomorrow.

In bad companies, very little thought is given to the whole leader vs. manager issue. The words are basically interchangeable, but this is misleading. There are real leaders in bad companies. However, unlike the "champions of change and innovation" in good companies, these leaders in bad companies are heading the charge to keep things the way they are. And since bad companies are so decentralized (something that will be presented in the next chapter), you can have these *effective* leaders at many levels within the company. These people are the enemy to any CEO who is trying to turn around a bad company. These *Terrorists* and *Rebels* (as they are called in later chapters) are generally not salvageable and need to leave the company as soon as possible. Only then is it feasible to work with and convert many of the managers who are left.

Who Needs This Book?

Virtually anyone who runs a company or a piece of a company can use this book. You may think your company is good. This book will help you determine that. If you are evaluating a company for potential acquisition or looking at the CEO position for a company with issues, this book will be a resource that will help you understand your target. Employees who are not sure about their company can get some guidance from this book, and it will assist job seekers in their evaluation of potential places of employment and help them ask the right questions. There have been times when we wished the board of directors of companies we worked at, or consulted for, had read a book like this. And bankers and stock analysts should be paying more attention to the cultural and operational aspects of the companies they cover. Financial reports do not always give you the whole story.

One Thing to Remember

A theme that is woven through this entire book is that the forces of dysfunction are inherently stronger and more natural than the forces that transform companies for the better. As an example, take the lawn at your house. If left to its own devices, you get weeds and bare ground. The beautiful lawn is the product of much investment, much hard work, and continuous attention. If the investment, the work, and the attention stop or are markedly decreased, the lawn quickly reverts to weeds and bare ground. Eventually, there is no evidence remaining that the lawn was ever beautiful. Companies operate in much the same manner. They naturally become dysfunctional when neglected and abused, and this all happens quickly. The positive transformation requires much investment, hard work, and continuous attention—and that change requires a much longer period. If you want to turn a bad company around, be cognizant of the massive commitment and also be aware that the change process is not a one-time thing. That commitment must go on forever or the company will quickly revert to weeds and bare ground.

Chapter 2

Identifying the Dysfunctional Company

Chuck: Dysfunctional companies, or companies that fail, are not the product of random circumstances. These companies all possess key characteristics that have been developed with great care over a period of time. Failure is not an accident. There are specific reasons for failure, and the process is surprisingly well planned and organized. If this is the case, one would ask if companies and their leadership intend to fail. While this is probably not true in most circumstances, fail they do because they: 1) do not know how to succeed; 2) make bad decisions due to ego, laziness, or habit; or 3) have priorities that, in their mind, define success but ultimately lead to failure. These key characteristics of failure can be very complex and have many variations, but for simplicity they can be distilled down to four categories:

- A culture of failure

- Ineffective leadership

- Poor strategy and processes

- No value placed on people

The rest of this chapter will investigate each of these character-istics in detail and provide examples so that the reader can easily identify a company in trouble or one that is well on the way.

Before we go further, it is necessary to present our definition of failure so the reader understands what we are trying to address. In our view, company failure involves performing at a level that is below what would be possible if a company had good leadership and management, employed strategies and processes that sup-ported good operational performance, cultivated a workforce that was of a good standard, and had a culture and values that could be considered good. In this case, *good* is not considered world-class but rather is thought to be at a level that a rational observer would consider to be an acceptable discharge of collective responsibility. In other words, the performance is not impressive nor is it seen as a squandering of company potential. The bases are being covered and there are reasonable results, but there is room for improve-ment and opportunities are being missed. However, it would take "best in class" people doing impressive things to attain that level of performance. When a company is in failure mode, rational people who are paying attention are not happy with what they see. There are easily identifiable flaws and gaps in people and the things they do and the decisions they make at the top and throughout the entire organization. A dysfunctional company could not survive an outside audit of any part of the company if that audit were performed by professionals not immersed in a culture of failure themselves.

A Culture of Failure

Company culture is a term that is often used and just as often not really understood. A colleague once said that "company cul-ture can be defined by what employees do when their boss isn't looking." It is really more complex than that, but this simple defi-nition provides a good start to understanding the massive impact that company culture has on what companies value and how they behave.

The culture within a company defines that company. A dysfunctional company has a dysfunctional culture. As an example, assume we have a company where the message from the top is to make as much money as you can. This message is the only thing the CEO and his senior staff talk about—all the time. They are totally focused on it. The best rewards (bonuses, promotions) go to the people who make the most money, no questions asked. Those who lag behind on this single metric are demoted, fired, or pressured to get better at making money. It is clear that the only thing that is valued is making money. Nothing else is important. On the surface, this looks like a fairly good plan, but it is a recipe for disaster. Employees start creating projects that yield great returns now but over the long term will implode. The finance team builds financing vehicles that actually generate short-term profits on their own but are incredibly expensive down the road. Acquisitions are made that basically do the same thing—profits on the front end and disaster on the back end. Accounting is pressured to find creative ways to increase earnings per share *now*. Ultimately, *now* turns into *later* and all the accumulated bad behavior brings the company down in a sea of failed projects and massive debt. Making money is obviously a good idea, but it has to be balanced by cultural values such as ethical behavior, making good economic decisions, and taking care of the long term. Dysfunctional companies do not understand this part of the cultural equation.

As the preceding example clearly illustrates, the culture within a company dictates the values and standards of behavior that the company and its employees embrace. In virtually all cases, the CEO and his team deliver the messages, either verbally or through their collective behavior, that define the company culture. They may or may not be aware of the company culture, but they have one, and they are communicating it either directly or indirectly to the rank and file every day. For companies in trouble or on their way to trouble, that culture is full of bad things that lead to very bad behavior. Sometimes a bad culture within a company has been around for so long and has ingrained itself in so many employees for virtually their entire careers that a new CEO may be unable to change things—unless he is willing to challenge everything that company is and values, and demand that total change is the only option for everyone. Bad company cultures do not need tweaks; they demand a complete overhaul.

The culture of a company in failure mode contains values that run counter to organization, progress, and success. Managers in these companies value independence. They would rather struggle on their own than accept input from anyone considered an outsider, such as corporate area experts or consultants brought in to help fix a specific problem. These same managers believe in the status quo. Change is always bad and to be resisted at all costs. In their view, it is better to fail with the old methods than succeed with something new. They are also totally focused on themselves and possibly their operation and do not have a company-wide view. As long as they can convince their superior that they are doing well (which may not really be the case), how the company is faring is of no importance to them. Processes, metrics, reviews, and strategies are all bad ideas in their opinion. All of that just represents extra work that they do not want to do. They want to be left alone, free to do what they want, and not to be questioned about anything. At the end of the year, they expect a bonus and a pat on the back and hopefully an easy budget for next year.

A big part of any dysfunctional culture is what can be termed "the myth of good." When managers at any level are asked how well their team performs something, the answer always is, "We're good." Whether they all actually believe it is open to debate, but it is obvious that they are not interested in exploring the subject in depth and possibly finding ways to get better. The myth of good is often actually code for "none of your business." This collective denial of reality can make its way even to the top of an organization, where the CEO will boast that his company is an industry leader when it is apparent to any rational being that there are major organizational and performance issues that are being ignored. Needless to say, a company full of people who "are good" is not interested in scorecards and quartiles, because they would actually determine who and what is "good." Amazingly, the managers who are better than average also subscribe to this attitude. They believe it is better to be part of the club than to call attention to their better performance or see the company do some things right for a change. Peer pressure is huge in a dysfunctional culture.

Another major characteristic of a dysfunctional culture is the acceptance of mediocrity. This goes hand in hand with the *myth of good*. People are mainly in a survival mode and have no expectation for progress or success. Whether they do not understand how to get better or find it all just too difficult to do, they have no appetite for it and are content just to be *good*. In some cases, new managers come in with ambitious plans for their operation but are beaten down by a culture that forbids change and will not allow progressive managers to make their peers look bad. These new managers finally get the message and either back off or move on to a company that will appreciate their efforts. People in a culture of failure believe the company will always be there and there is no real reason to get better. There is no urgency or any feeling of ownership or accountability. When no one yells "Stop!" wrong behavior and actions become the accepted norm.

Everyone in a dysfunctional company, from top to bottom, has an abhorrence for accountability. Top-performing companies like to talk about clarity in goals, results, feedback, and expectations. In bad companies, they want to keep all of those items murky and definitely not on the list of topics for conversation. A scorecard manager once told of being chastised by an SVP who was responsible for more than $3 billion in revenue because his scorecard and quartile results were hurting the feelings of underperforming managers. The idea that these results could be the start of some very productive conversations about improvement was totally lost on the SVP.

Both managers and their direct reports hate goals, goal setting, reviews, and evaluations. It is easier to keep everyone equal and avoid having to give negative feedback that would upset an employee and cause some conflict. That person may resign and then would have to be replaced, which is just more work. Managers do not view goal setting as the foundation for performance management but rather as a task they are not trained to do and therefore are not comfortable doing. They tend to wait till the last minute and then put minimal effort into it. A typical example is when the president of a large service company barged into the office of a direct report and frantically asked for some ideas about goals for all the vice presidents. He had to give the board of directors these goals in an hour and had nothing. He viewed these goals as just another unpleasant task and not as the foundation for progress for the next year.

Line managers and other employees likewise do not like goals and metrics or anything that can give anyone a clear view of their performance. In the absence of specific information on performance, they can use their expertise at excuse making. Their experience tells them that if they answer enough questions with good excuses, the questioner will lose interest and move on. If the questioner persists, the trump card is to respond that their situation is unique and that their performance cannot be compared to that of anyone else. Since the questioner usually has only sketchy information and is uncomfortable doing any kind of review, this strategy usually meets with success.

In this environment where there is no discipline or accountability, fear of the truth and its inherent responsibility is paramount. People just want to do their jobs their way and be left alone. They do not want to commit to anything or be told their performance does not meet standards and they have to improve. Success in the bad company is mainly based on maintaining a friendship with their superior. Managers are equally happy with this arrangement. They do not have to do anything that is uncomfortable, such as dealing with noncompliance, and they can take care of their friends with bonuses and advancement. Furthermore, they do not have to justify anything because the facts have been suppressed.

Bad companies always have ethics issues. The practice of good ethical conduct requires standards, accountability, taking responsibility, and making a commitment to doing the right thing—even if it is not in your best interest. You have to value doing things right. In fact, doing your job to the best of your ability is an ethical issue. The culture of a dysfunctional company does not contain these values and these priorities. Consequently, ethics is rarely addressed, almost never enforced, and is generally viewed as a nuisance. The CEO may make grand public statements from time to time about the importance of ethical behavior, but it is just a hypocritical show for those watching. The rank and file understand the real message, which is that ethics is only a political topic and not anything of real importance. The most telling comment was made by the president of a large public company who remarked, after delivering a speech on the importance of ethics, that ethical behavior often got in the way of creativity. That company later got into trouble because of excessive creativity and the president lost his job. When ethics is not a key value in the company culture, the company and its people become untethered. The good people grow frustrated and leave, the bad people thrive, at least temporarily, and the fence-sitters become bad people by default. It is interesting that when bad companies finally implode and the investigations begin, there are always many instances of employee theft and self-enrichment. This should not really surprise us. When you hire people to steal *for* the company, they will ultimately steal *from* the company.

This section on culture has not totally covered everything that one must know about the topic, but it has presented a good snapshot of what the culture looks like in a company in trouble and the role that bad culture can play in creating the damage that will ultimately destroy that company. Later in this book, there is a whole chapter on proper company culture, what it contains, and how it can be developed and sustained. The key messages to remember here are that culture is very important, is a great indicator of the health of a company, and never lies about what it represents. When evaluating a company, people are well advised to start by taking a hard look at the culture that the company has embraced.

Ineffective Leadership

We saw in the previous section that company leadership plays a major role in defining and creating the culture of the company. Poor leaders with bad values produce cultures that are highly damaging. However, leadership flaws can go beyond the cultural aspects in having very negative impacts on how a company and its key people behave and function. When leaders are not effective, they directly damage everything they touch. In this section, we will explore how bad senior leaders affect the various levels of management, make it difficult to get things done in the company, consistently stymie progress, and dramatically add to the dysfunction.

Let us start at the top with the CEO. In the American business environment, the CEO is king. He is seen as the face of a company, the only one with good ideas, the one who makes things happen, and the one who gets the lion's share of the credit and compensation when results are good. The media business shows treat CEOs like rock stars, or at least like celebrities. As Mel Brooks would say, "It is good to be the king." On the positive side, a good CEO can make a massive difference. Steve Jobs and Jack Welch were a big part of the identities of the companies they led. Wall Street attached great value to their presence at the helm. By the same token, in dysfunctional companies, the CEO plays an equally key role in bringing about the chaos and the failure.

Before we get into the specifics of how a CEO can ruin a company, let's establish a positive benchmark so we know what is missing when there is bad leadership and a company is in trouble. At a minimum, a CEO must be seen by the rank and file as a capable leader who can inspire action. He must assemble a talented team and then use that team for advice and support to do what is best for the company. The CEO must always have a current realistic view of what is happening in the company and in the marketplace so that he can mobilize the company to properly respond. In short, an effective CEO must be totally connected to the company and then use all the assets that are available to make the company successful now and over the long term. Any CEO attitudes or activities that run counter to these requirements will reflect negatively on the company.

What follows in the next several paragraphs are descriptions of attitudes, behaviors, and practices that can lead the CEO and his company down the road to disaster. The list of failure mechanisms is not totally comprehensive, but the reader should get a fairly clear picture of how a bad CEO can tear the heart out of a company and basically paralyze it. We will start with a very common

19

malady that we call the "CEO disease." Virtually every CEO eventually is afflicted with this disease, because the human ego and the people who feed it are ever present. The *CEO disease* need not be fatal to a company if the affliction is light or if the CEO does not allow it to significantly compromise his performance or that of the company. Companies can still flourish if the CEO has the disease under control. However, for many CEOs, the disease gets in the way of everything to the detriment of the company and its employees. It is probably safe to assume that most CEOs start out with the best of intentions. They may even be a little intimidated by the huge responsibilities of that first CEO job. In fact, a recent (2018) survey of 402 international CEOs showed that 68 percent of those respondents did not feel they had been fully prepared for the job. Their knowledge base concerning what the company does and how it operates may be light, or at least not complete, and they know they will have to depend on others who have been around longer to help them understand issues. They intend to work with people, take advantage of good input from the people around them, make good decisions in the right way, and lead the company to success. Somewhere down the road that all changes. It can take six months or several years, but eventually, after being told how smart and insightful they are by so many people, they begin to believe it. They become not only the "smartest guy in the room," as might have been heard at Enron, but often the only smart guy in the room. What happens next is the onset and development of the CEO disease.

CEOs seriously afflicted with the disease believe they have God-given instincts that make them right about everything. Their opinion is the only one needed, and dissenting views are not tolerated. In fact, most people disagreeing with the CEO are putting their careers in jeopardy. Meetings with the CEO have little dialogue but mainly edicts from the CEO and fawning compliments from the masses. In one case, two VPs working on a cutting-edge efficiency initiative that was in its initial stages met with the CEO to brief him on their progress. Their intent was to communicate their cautious optimism but to detail the many technical hurdles with no easy remedies that were still in the way. Computer code that did not exist needed to be created and there were instrumentation issues that suppliers indicated would take at least six months to get right. In short, there was still a lot to do before they could scale up beyond the pilot phase. About ten minutes into the presentation, the CEO cut them off with, "I do not want to hear any of that. I have my own plan, which is to implement the whole thing now." The two VPs left the meeting knowing their project had just been destroyed and that the next year would be wasted trying to get a half-baked program at least to accomplish something. What started out as a very promising efficiency initiative that could have changed the company eventually failed miserably because the pieces never came together and functioned properly. Too many bad decisions and equally bad commitments to suppliers were made in haste, way before the necessary information was available to enable the proper decision making. They guessed and they guessed wrong. In the end, the greatest damage was from the anger and lack of trust generated out in the field. Line managers had always resented corporate people dabbling in their operations and creating more work for them. This just added more fuel to the fire.

CEOs with the disease create their own reality. When you have high confidence in your own observational skills and you have eliminated everyone around you who might offer another perspective, then your view of things becomes the truth. Plus, everyone is always telling you that you are right. In this environment, senior managers who know better stare at the floor as the CEO pontificates on how things are and what will be done next. They know these plans are doomed, but it is pointless and even dangerous to raise the issue. A CEO once assembled his senior staff to work on strategy. He started the meeting by stating that he was the only one who had ever had any ideas and that the rest of the people in the room were worthless. Of course, he rarely let his staff members talk and was often quick to reject their ideas out of hand. After the meeting, which was not very productive, some of the attendees compared notes on whether the CEO had ever really had an idea of his own. His normal practice was to adopt the ideas of others, and the few successful projects that followed, as his own and then take credit for the results. Needless to say, the trust level between this CEO and his staff was not very good. CEOs who behave in this manner have effectively eliminated the benefit that comes from having a staff around them. They still have the cost but not the positive impact. As children, we learn about blindness caused by vanity in stories such as *The Emperor's New Clothes.* But as adults, some people have forgotten those lessons.

If CEOs with an advanced version of the disease stay long enough at their company, they end up isolating themselves from all but a few of their favorite fawning lackeys. People avoid the CEO and do not want him in meetings because his presence just sucks the air out of the room. Interactions are dysfunctional and uncomfortable. It is hard to have a dialogue with someone when the conversation is in only one direction. People tire of the abuse, and that drives them away. No one wants to bring the CEO any information, especially if it is negative, as it may be poorly received and they do not want to be the messenger who gets killed. The CEO is generally fine with all of this, since employees do not merit his time anyway. As the CEO becomes more isolated and bloated with ego, he has more time to concentrate on what is most important, which is what the CEO is getting out of the company. Since the company could not survive without him, he feels he is entitled to more. What follows are huge deferred compensation packages, numerous perks, extended family vacations using the company jet, and a lot of time away from the office. The senior staff is just fine with the CEO's time away from the office because they can actually get something done then.

We earlier said that good leaders inspire action. CEOs with the disease inspire distrust and inaction. It is like having a quarterback in the huddle who is playing for the other team. People are paralyzed by the state of things and so is the company. There is no chance for anything good to grow out of such a dysfunctional situation. Can a CEO with the disease be remediated? It is not impossible, but how do you make people who are not open to input be open to input about their own behavior? Good luck!

Although the CEO disease is responsible for much of CEO failure, there are other ways for a CEO to be an obstacle to the success of the company. As we said earlier, anything is on the table that prevents a CEO from being connected to the company and its business—having an unrealistic view of things, not making good decisions based on the best available information, and not leading and inspiring the employee base to put company plans into action over the short and long term. What follows are a number of other causes of CEO failure. The list is by no means totally comprehensive, but it covers much of the topic:

- Poor people connection—Some people just cannot connect with others. They may be bright, they may work very hard and always be prepared, and they may have great ideas, but when they walk into a room, everyone else wants to leave. It is hard for these people to rally the troops and get people moving on important things that should be done. It is easy for employees to view these people negatively, and much of that negative energy gets transferred to the company. Unless these people can find ways to compensate by getting some coaching and surrounding themselves with trusted lieutenants who can get the message out and engage the workforce, they are doomed to failure.

- Bad standards for people evaluation—CEOs can create failure when they are blind to the actual ability and behavior of the people they hire and employ. This is often a symptom of the CEO disease. CEOs with this flaw consistently hire bad people and then become oblivious to their poor performance and unacceptable actions. Senior managers who have not been totally corrupted by the dysfunction are horrified when a key position is vacated, as they wonder just how bad the new "key person" will be. In good companies, these are the strongest people in the organization and they are tasked with some of the most critical responsibilities. In bad companies, these inept new leaders collect more bad people around them, inundate the company with unacceptable policies and standards, destroy company morale, and derail anything good that may be happening. With the broad reach that a senior position offers, these bad people can do an awful lot of damage. This reflects directly on the CEO. Good people who see nothing but bad management above them quickly move on so that more bad people can fill the empty seats. Eventually, the trickle-down process is complete as the bad senior team creates the bad company. That may not be what the CEO had in mind, but that is what he will get if he is not aware that he needs a lot of help in the personnel area.

- <u>Weak decision-making skills</u>—Some CEOs are natural "hip shooters" when it comes time to make an important decision. This also can be the product of the CEO disease. Even though they may have access to the right information and analysis, and the counsel of smart people who may be closer to the subject or have more knowledge or experience, they prefer to "go with their gut" and make a quick, personal decision and then move on to something else. This practice, which depends too much on luck, is going to produce some horrible decisions that can doom the company. CEOs who know they tend to get impatient and lazy when the hard decisions are in front of them need to surround themselves with robust decision-making processes administered by a strong team of respected senior people who can stand up to the CEO. Furthermore, the prudent CEO will consistently reach out to the board of directors for oversight input to make sure that the resulting decisions consider all the available key information and can pass the reasonable scrutiny test.

- <u>Inability to prioritize</u>—This issue is typically reserved for those CEOs who are incredibly bright, energetic, and creative. They are "on" pretty much 24/7, think about the company all of the time, see opportunities way before anyone else, and have the insatiable need to drive those opportunities to fruition. Companies can thrive with this kind of insight and drive, but when the "game changes" every fifteen minutes, the senior management team is consumed with rushing from one priority to the next and nothing ever gets finished—or at least finished correctly. Everyone is working hard and the energy level is sky-high, but real progress is hard to find. Key people who bear the brunt of this kind of chaos eventually burn out and move on. Prudent CEOs need to continuously take stock of what is happening in the ranks. Messages and priorities will change, but they must be managed, and those on the receiving end need to feel like they have a chance to win and are not just hamsters in a wheel that moves faster and faster.

- <u>Self-interest</u>—This is a special case that we have only personally seen once, but we believe that it is more prevalent than that. Here again, it is part of the CEO disease. In this case, the CEO appears to have all the tools to do the job. This person is bright, hardworking, and insightful; really understands the business; and has the ability to inspire people and lead. The problem is that the CEO views the company as his personal vehicle to success. What he wants, such as personal enrichment and power, takes precedence over the interests of the company. The CEO may make bad acquisitions that damage the company but fulfill some personal ambition to run the largest company in the industry. Or he may make vendor or customer commitments that are not good for the company but are personally rewarding to himself. A CEO cannot be totally effective when every decision revolves around "What is in it for me?" These CEOs can end up with ethics issues involving side deals, kickbacks, or special stock deals that put them in the newspaper and bring their career to an end. Unfortunately, the company they run has to suffer in the process.

Many people think that all CEOs are alpha male or female extroverts who are highly intelligent, have tremendous energy, are strategic geniuses in their industry, and are natural leaders. The media does a good job of reinforcing this image. Reality presents a different picture. CEOs come in all shapes and sizes, and there is no template of the skills needed to get the job done. We have seen introverts, people of average intelligence, and people with marginal work habits in the CEO job. Some have done well and some have failed. The key difference is that those who succeed have the ability to take stock of what they bring to the table and then compensate for their shortcomings by building the needed capabilities into their personal tool kit and/or surrounding themselves with the right people and processes to make them effective and keep them out of trouble. This all takes a lot of self-assessment and reality checks, and the ability to accept and listen to input from others. In the dysfunctional company that rarely happens, because all of these activities are just too uncomfortable and too much work. Whether a CEO is afflicted with the CEO disease or just does not have all the skills, attitude, or perspective to do the job, when these issues are not properly addressed and the CEO position is compromised, it is hard to do much more with a company than keep it running at a low level. If there are big problems that require big changes, that is a bridge too far for a company that is effectively without a leader. As we will see in later chapters, companies can only tackle their biggest problems and make big changes when the leadership is not only good but outstanding and very tenacious. All good companies have this going for them. Bad companies do not unless they do something about it.

There are other senior executive people in a bad company who can do nearly as much damage as the CEO. In fact, in most cases where a new CEO comes on board to remediate a company in trouble, they have to deal with and are often brought down by the next layer of management. The area presidents and SVPs (AP/SVP) who are directly responsible for big pieces of the company revenue are virtual CEOs to their people. To many employees, in their part of the company, they are the face of the company. If a CEO is going to be successful, he has to really connect with these people, rely on them, and function with them as a team. Conversely, when there is a disconnect between the CEO and an AP/SVP, the piece of the company that is controlled by the AP/SVP is virtually cut off from the CEO, and anything that comes out of the corporate office never gets there. That happens a lot in dysfunctional companies. A CEO may have some really good ideas for improving the company, but if the AP/SVPs are not buying in, then none of it happens. The AP/SVPs are typically more subtle than just refusing to do it. They may nod and smile and say they are all for it—but it never happens. There are myriad excuses, which usually revolve around the argument that the initiative is not well crafted or just not feasible, and after a while, the CEO gives up and tries something else. The big question is what is the motivation for this kind of obstructionist behavior? For many dysfunctional AP/SVPs their part of the company is viewed by them as their personal fiefdom. They want total control and total allegiance from their people. They resent the CEO, let alone anyone else, dabbling in their territory. Even if the dabbling could produce some benefit, that is of little importance. Everyone must stay out. An example in a large service company with more than 400 offices involves the director of a significant improvement initiative. This person was experiencing major resistance from an SVP and getting nowhere with that part of the company until one district manager agreed to try the program. The district manager had resented being ranked below average by the director, even though his financial results were on budget. The director argued that the market was inherently lucrative and the district manager was not coming close to tapping its potential. All the nonfinancial metrics verified this. In frustration, the district manager decided to give it a try and increased his margins by 50 percent, along with positive improvements in safety, turnover, productivity, and operating expenses. At a company meeting for all managers, the district manager was one of the featured speakers because of his performance improvement. Onstage he told his story, praised the program, and thanked the director for his tenacity. He said he no longer was fighting fires all the time. He was now proactive instead of reactive and had time to really manage his operation. The director thought, "Finally, a breakthrough." But then he looked over at the SVP, who was seething with anger. If looks could kill, the district manager would have been dead on the stage. At that point, the director knew the war was far from over.

The big question is, how is this allowed to happen? In some cases, the CEO may not be paying attention, but it is hard to believe he does not know. Many CEOs are simply afraid to make it an issue. These AP/SVPs are seen as valuable people who are connected to huge chunks of revenue. The fear is that their removal could damage the company and it would be hard to replace them. So their bad behavior is tolerated. The late Michael Hammer, the process expert from MIT, used to say that bright and productive people who were not onboard with company plans are the most dangerous people in the company. The damage they do far exceeds the benefit they bring. However, CEOs in dysfunctional companies rarely come to that conclusion and do something about it. Rogue AP/SVPs are free to sabotage any efforts to improve the company, and their track record is impressive.

Dysfunctional AP/SVPs are very aware of their impact on the company. Like the CEO, they are often told how brilliant they are. They feel they are more important than the CEO, who they see as naïve and really out of touch with how things actually run. Many of them believe they should be the CEO. The big motivators here are ego and control. The same SVP referenced previously brought his managers in to plan the coming year. The CEO and his corporate team came in for the first day of a two-day session to give the company perspective and go over what they wanted from the SVP's team. After what appeared to be a very positive first day, the corporate team left. The SVP stood in front of his people and said, "We are not doing any of that. Those guys don't have a clue. I will handle the CEO. Remember, you work for me. I am the only thing between you and them. I am protecting you." This tactic does several positive things for the SVP. He becomes the most important person to all his people. He has them all united in their hate and distrust of the CEO and the corporate team, and they are now beholden to him because he is protecting them. Unfortunately, the price for this is very high. He has effectively marginalized his part of the company and cut them out of any chance of taking advantage of anything good that may come out of the corporate office or any other part of the company. The entire company gets to suffer for his bad behavior.

AP/SVPs do not take the CEO on directly, but they know that if several of them band together, they can wield significant power and ultimately get their way. In fact, if enough of them unite and the CEO cannot or will not stand up to them, they can achieve critical mass. In this state they effectively run the company. The CEO may think he is in charge, but the AP/SVPs are determining policy and defining the culture. They can isolate the corporate office and the CEO from the rest of the company, build an operating culture where wrong actions are never challenged at the operational level, create a general fear or lack of desire to deal with issues, and make sure there is absolutely no employee identity with the company or feeling of ownership. The company ends up looking like Afghanistan. The president may think he is in charge, but he is effectively the mayor of Kabul. The various warlords run the country. In this setting, the CEO does not want to do battle with the AP/SVPs. It is easier to give them their way and keep them happy. And what does giving them their way and keeping them happy actually involve? They basically want to be left alone, free to operate their piece of the company as they see fit. They want unlimited access to capital and they want bonuses and other perks for themselves and their direct reports independent of how the company does. They do not want to be evaluated, ranked, reviewed, or put on improvement plans. When a company gets to this state, it no longer functions as a company but as a collection of smaller entities that operate independently and make little or no progress.

One may ask why these smaller entities cannot be well-run. The answer is actually fairly simple: When a leader, in this case the AP/SVP, unites his people by rejecting authority and embracing bad behavior, the culture that is produced cannot be anything but dysfunctional. These people do what they do because their values and priorities are wrong and the organization that they produce is full of bad people who likewise think and behave badly. Chaos breeds additional chaos. In their own minds, they may have the best of intentions, but it is their behavior, the decisions they make, and the messages they send to their people that will determine results. They can negatively affect virtually everything in their purview and everyone below them on the organizational chart. Bad leaders tend to generate a lot of bad management around themselves. The reasons are many and include people tending to hire others with similar values and styles or rewarding bad behavior and therefore getting that kind of behavior in their direct reports. So the mayhem migrates even deeper into the organization and negatively affects more people and more of the business. If left unchecked, a dysfunctional AP/SVP can totally ruin his piece of the company.

This section has focused on the CEO and the next layer of management, the AP/SVP. However, these observations can apply to anyone who has authority and influence over an office, a district, a division, or any other piece of the company and its employees. The process of dysfunction and failure is basically the same at all levels. If no one is paying attention, dysfunction moves quickly and destroys everything good in its path. The system of leaders/managers in a company is much like the circulatory system in the human body. When a leader or manager becomes an empty chair, or worse, that part of the company gets into trouble and becomes sick. And like the human body, sickness tends to spread if left unchecked. Unless there are timely diagnosis and intelligent intervention, the patient can ultimately die.

Poor Strategy and Processes

One given that you can pretty much take to the bank is that good companies are good at strategy. From the CEO to people well down in the organization, they understand that strategy is the game plan for the company. It defines where the focus and the energy are directed, and it unites everyone in a common effort to outmaneuver the competition and succeed along the path created by the company's strategy. Senior leaders in good companies are well versed in creating the right strategy, uniting their people behind the strategy, and then putting the strategy into motion and finding success with it. If you do not understand any of this, are not comfortable with it, or think it is all a waste of time, you will not last long in a good company.

Bad companies typically are mystified by the whole business of strategy. The leaders and managers in these companies have gotten through their careers so far without it and do not understand why it is needed. In their minds, strategy is what business consultants are always touting to naïve CEOs as a way to get more consulting work. A senior leader in a large corporation walked into a meeting the CEO had called for senior executives. He was early and asked the other two early arrivers what the meeting topic was going to be. One of them responded that, "The CEO, for some reason, has gotten into this strategy BS. It is going to be a waste of time." Given these initial attitudes, it probably was a waste of time.

Development and implementation of strategy requires discipline, an in-depth understanding of the business, creativity, the desire to change and grow, and the ability to all work together for a common goal. We know from the previous sections in this chapter that dysfunctional companies are not good at any of these things. Strategy building requires the company leaders and managers to collectively take an honest look at themselves, their skill sets, and performance gaps, how they stack up against their competition, and what is needed to be successful in the marketplace going forward. When your culture avoids self-assessment and likes to keep the facts murky, all of this is very uncomfortable and something to be avoided. The president of a large company called in a direct report and proudly announced that they were going to acquire the largest company in their industry. The direct report was mortified. He responded that the company to be acquired had struggled for the past ten years because they had outgrown their strategy and could not change. He said, "Why are we the solution? We have the same strategy and behave in the same way. This will all end badly." The president looked puzzled and just responded that it would work out because "We are better." The whole company failed almost exactly a year after the acquisition closed and the president, along with 90 percent of the senior executives, lost their jobs. The direct report did not lose his job and got involved in the rebuild.

Bad companies do not have a successful legacy with strategy. Since leaders tend to collect people around them who have similar histories and views, bad companies generally collect people who have had bad experiences with attempts at strategy. In their collective minds, it is something that other people in other companies do for strange reasons. It does not work for them and they feel it is not necessary. They have gotten through life just fine without it. It may happen that a new CEO comes on the scene or a consultant catches the ear of the existing CEO and, all of a sudden, there is a desire to create a strategy. For some of the people, this is just a newer version of the same old horror story. They are negative before the effort even starts. Some people may actually get excited. This may be their opportunity to get involved in something that is first-class and sophisticated like they do at world-class companies. Maybe this idea will really work for the company and they will finally understand what this strategy business is really all about. They hope the mystery will be solved. These initiatives start out with great enthusiasm and fanfare, at least for some of the people. Teams are formed and the CEO and his consultant make speeches about becoming a great company. There are breakout sessions, planning retreats, and more consultants are brought in to help with specific tasks. At some point in the process, usually when planning documents are in draft form or possibly near the beginning of the actual implementation phase, things start to unravel. Managers start grousing that their people assigned to the strategy initiative are not getting their "real" jobs done, the naysayers complain that the plans are not feasible and will actually damage existing programs, and the team members become frustrated with all the pushback and negative input. Then the compromises start with the scope cut back to accommodate the naysayers, schedules start to slip, and managers push for exemptions. The core group responsible for the initiative were once heroes at the beginning of the project but are now pariahs. Eventually, the strategy initiative just fades away. There are no announcements, there is no wake or funeral; people just stop talking about it and the team members go back to their original jobs. The naysayers will smile amongst themselves and tell their people, "I told you so." The CEO will go back into his office and rationalize the failure, and the consultant will be paid and will start thinking up a new project. The people who were once excited are disappointed. They always suspected that this whole strategy thing was not for them and the company, but they had hoped that it was. Now it just continues to be a mystery.

Eventually, the company will make another attempt at developing another strategy and it will also fail. The reasons for failure have all been previously discussed in this chapter. Strategy was not a value in the company culture and therefore was not seen as vitally important. It was more of a fad to most, and maybe a silver bullet. The CEO had not neutralized all the obstructionist forces in the company that try to sabotage every attempt at change, and he allowed those forces to take over in the trenches. For strategy to work, the CEO has to take ownership and communicate to everyone that failure is not an option. In this case, failure and then going back to the status quo was always too easy an option, and eventually everyone took that route. Building and implementing strategy is hard work. It takes commitment on the front end and tenacity all the way through. People need to know that *on day one* and understand what is on the line and what is expected of them. Noncompliance and bad behavior must be dealt with quickly and publicly. You have to set standards to make strategy work, and bad companies are not big on standards.

Strategies are a must for companies that are progressive, forward-thinking, and at the top of their industry. However, for companies that are stagnant and mired in simply doing what they do, year in and year out, independent of the marketplace, you can understand but not accept their dislike of strategy. However, you would think that they have to at least have processes to be good at what they do. Amazingly, the bad company people are as negatively aligned on processes as they are on strategies. Here again—processes, like strategies, demand commitment, accountability, teamwork, and management of what everyone is trying to do. People in bad companies do not like any of this. They want to be left alone to do things their way or whatever way they feel like doing them on any given day. Processes do not let you do that.

Within a process, there is a standard way of doing things. Those procedures were adopted because they are the most effective and most efficient. They may require more attention, more documentation, more reporting, and more management, but they work when everyone does their piece. The problem is that in a dysfunctional culture where everyone is *good* and *better* is not valued, few people feel that processes are justified. They are just too complicated and require too much discipline and work. Plus, you have to train people, which is really a nuisance and a time waster. A senior manager in a large service company once observed that his people were "consistently inconsistent." That observation was absolutely on the mark. Unfortunately, that was the end of the discussion. The manager saw it only as an observation, the way things were, and there was no reason to do anything about it. To that manager, processes either were not the answer or were a bridge too far for his operation.

Sometimes process ignorance is the reason nothing is done. Managers may know they have a problem, but they have no idea what to do next and do not feel the need to ask. That may involve admitting they do not know something or asking a corporate person or a peer for help. To these managers, neither of those actions is acceptable when one has an image to maintain amongst one's peers. It is easier or more politically correct in a dysfunctional environment to take the position that processes are a waste of time and not needed here.

In a company without processes, you may find a manager who has basically built his own set of processes. Most of these processes do not qualify for best-practice status, but sometimes they can be surprisingly good. The processes are underground and do not get publicized. The local manager knows he is swimming against the cultural current and is just as happy to keep things quiet. The sole motivation is getting the job done, and these managers just happen to be more enlightened than their bosses. Hopefully they will eventually move on to a better company that will appreciate their efforts.

In bad companies, efforts to build new processes or keep the few old processes going are squashed pretty much the same way the strategy initiative was torpedoed earlier in this section. The resisters to change line up and take control, and the CEO backs off. If the resisters cannot stop a process, then they will make sure that it is not properly managed. A service company developed a new productivity process when it was determined that managers were not setting targets for drivers, were not doing ride-alongs to calibrate routes, and were not holding productivity meetings with drivers to discuss performance or ways to get better and deal with issues. After some period of time, it became apparent that the majority of managers were not managing the process. There were targets for each driver, but no one could explain how the targets were set. Reports looked like they had been fabricated, as every driver was meeting his target. There were verbal reports from insiders that ride-alongs and meetings were not taking place. The director in charge of the process went to the president and described the problem. A few weeks later, at a meeting with several of the more difficult managers, the president asked that "they get better at productivity." Needless to say, this conversation had no positive impact on the issue. The noncompliant managers knew they had a president who lacked the knowledge, interest, or fortitude to set them straight, so they continued to do things their way.

Processes are fabulous tools. They help new managers start out on the right path by doing things correctly at the beginning. They provide a great basis for training and they establish the universal definition for quality, and the right way to do things. When there is a prescribed method, the management aspect is a natural next step. Processes beg to be managed and offer great clarity on performance. Unfortunately, in bad companies, none of this is valued. In fact, most bad managers act as if they are allergic to processes. Look at a bad company and chances are you will find a company devoid of functional processes.

An extension of no strategy and no processes is a common practice that can be called the "office game." It is basically the way that managers and their support staff in bad companies operate. They appear to be busy and involved in many things, but they are basically busy doing nothing or at least busy doing only the fun and easy things. There are a lot of meetings with a lot of attendees.

Lunch is often served. Managers usually fill their whole day with meetings. Unfortunately, people leave meetings with nothing resolved and often throw the meeting notes away. The important things, such as setting goals, doing reviews, building processes, dealing with performance issues, and completing useful projects are not done because managers are just too busy with their meetings. In fact, "I am too busy" in this environment is actually code for "I do not want to do it." There are a lot of planning meetings, but little actual planning takes place. Really big problems can be buried by inviting as many people as you can to a series of meetings. When a CEO or other senior leader asks for a progress report, the standard response is, "We are working on it. Had a meeting with twenty-five attendees yesterday." The expectation is that something else will bubble to the surface and everyone who is interested will be distracted and forget about the issue. This kind of organizational cholesterol is effective at killing almost anything, including strategy, processes, and any attempt to change or improve things.

Sometimes significant issues do get in the way of all the fun and cannot be buried by meetings alone. In these cases, it is best to ignore the reasons for the issue and not attempt to find a solution. Reasons and solutions can lead to accountability and change—which are not good things. Instead, just create a new rule. This rule may not address the problem and may even cause more damage than the problem itself, but it is a response and it seems to satisfy everyone. The key is to appear busy and appear to be responsive but do as little as possible, do not take ownership of anything, or offer an opinion that is unpopular or isolates you from the crowd. It is all form over function, but it is the way to survive in a dysfunctional environment. No senior manager will ever ask direct reports what they have accomplished in the past quarter. A truthful answer would require action, and that is not a desired outcome.

The very small amount of actual problem-solving that does take place, usually by accident, is not high-quality. The participants are naturally not comfortable with having to do real work. They are not good at problem-solving, because they rarely do it and no one has ever shown them how it should be done. So everyone is faking it. Diving into real issues and evaluating the actions taken just runs counter to the way things are naturally done. There is too much reality here, and friends could get damaged. This all could trigger more issues and result in unwanted change. They believe it is best not to dig too deep, minimize the facts, and quickly arrive at a solution that is politically correct and does not bruise any feelings. One other option is to blame it on an unwanted process or a manager who is actually pushing for change and improvement. This kind of lazy problem-solving does not solve the problem, but everyone can now claim credit and go back to ignoring the problem and playing the office game.

The opposition to strategy or process or playing the office game is all based on a culture that does not value action, accountability, or results. People are more interested in just getting along and surviving without having to stand up for anything or make a commitment to improve themselves or the company that employs them. They are experts at rationalizing ineffectiveness and failure and feel no real ownership for their job or how they should be doing it. When everyone is on board with this approach, it becomes the accepted practice, and anyone asking for more is just a troublemaker.

No Value Placed on People

Management 101 tells us that companies are only as good as their leaders, managers, and employees. That is a pretty simple concept and is easy to understand. For some strange reason, bad companies have never understood any of this. Good companies, by contrast, know that attracting, training, and developing good people is a core competency that will differentiate them from their competition. If they have better people with better skill sets who are more motivated to be successful, they will come out on top. They will be more creative, more productive, more nimble, and more responsive to their customers because they simply have better people. That is how you win. When you work with what is left over, that is how you lose.

The first step in the people process is building the team. Dysfunctional companies never really build the team. They simply take who is available, who is already there, or who walks in the door. There is no quality process for building the team. They have no standards or any real idea what they should want. Like strategy and process, the managers do not understand it, are not good at it, and do not like to do it. Putting together job descriptions, interviewing candidates, selecting the new hire, and then breaking that person into the system are maybe even more painful activities for the manager than strategy and processes. They have never been trained to do these things, have never seen anyone do them well, and just want to get it all over with. Sometimes they feel that leaving the position open would be preferable to going through all the hassle. It is often easier to just call up someone they know from a previous job, even if that person is not very good. At least they are a known entity, they know the routine, and you can get everything over quickly. The HR department is of little help here. They see themselves as a support function that is mainly focused on benefits. They may help out with the negotiations with the new hire and maybe participate in the interview process, but that is about it. They are no help in setting up standards for positions or finding reliable sources of qualified people and then creating a pipeline of vetted candidates who have an interest in the company. Bad companies have such an aversion to the hiring process that it leads to a collective lack of confidence in their ability to fill key positions. A senior operations manager in a large company was once criticized for the overall poor performance of the managers under him. He responded that he would like to replace at least twenty-three of them but did not think he could find replacements, let alone replacements of better quality. He was more comfortable choosing known failure over a chance to improve.

In bad companies, the hiring process is always reactive. It only happens when people leave or a new position is created for some reason. Then the attitude is to just get it filled as soon as possible and end the fire drill. There is never any thought given to force-ranking employees and replacing underperformers or upgrading the talent in a key position. No one seems to focus on the fact that more talented people can do more good for the company. It may be that managers in bad companies innately know they are bad and feel that they do not deserve, nor could they keep, good people. It is easier to keep mediocre people and then, when the need arises, to hire more mediocre people. These employees have low expectations and will not be appalled by the lack of strategy and processes and the bizarre behavior that is always in evidence within the company. They will hang around and not have to be replaced.

Bad managers tend to hire people who behave just like they do and are perhaps a step or two behind on intelligence and ability. A senior manager once remarked that he liked to hire dumb people because they never questioned anything and would do everything he asked them to do. He built a team of very loyal employees, at least to him, who were not long on talent but were capable of just about any kind of unacceptable behavior. None of this bodes well for the company. But if you do not value talented people and put little effort or scrutiny toward whom you bring in to represent the company, then you cannot expect much in the way of results.

Once you bring someone on board, independent of their abilities, it is mandatory that you prepare them for the job they were hired to do. Bad companies do not do this either. They basically believe you teach someone to swim by throwing them in the pool. Training may be discussed but is really seen as a nuisance, just extra work and cost with little expected benefit. If there is some training, it is usually quickly cobbled together, of short duration, and not of much value. Like everything else, bad companies are not good at training either. New people—in the absence of training, on-the-job mentoring, or a job manual—usually get their job done by faking it. They try to stay busy doing what they know how to do. For instance, an accountant brought on as a new operations manager will spend his time in the office going over the books. A person with a sales background will concentrate on customers. The rest of the job goes begging. Eventually, if the employee is diligent and puts in the extra effort, he will develop some familiarity with the rest of his responsibilities. However, this is rare and does not give him the depth of knowledge that good training, on-the-job mentoring and job manuals would provide. The final product is a group of substandard managers who are only focusing on a piece of their responsibilities, while the rest is on automatic pilot. If the managers are not particularly bright or talented to begin with, the result can be even more dismal.

Once people are brought into the company and become familiar with their job, their performance needs to be managed. Bad companies do not do this either. Performance management is all about process, accountability, measuring results, and providing feedback. As we have already seen, this all runs counter to the culture in a dysfunctional company. Managers do not know where to start, so they do as little of it as they can get away with. Goal-setting is lazy to nonexistent, and performance feedback is spotty and not to the point. Managers in bad companies have been known to bring employees in to discuss goals and performance and then get distracted with conversation about hobbies or vacations. When time is up, the employee leaves and everyone is happy. As we have seen, it is easier for a manager to just tell everyone they are good and then give the rewards and the promotions to the ones he likes. Performance is not really valued, so there is no real reason to go to the work of managing it.

Since the company does not employ any means of measurement or assessment, no one is aware that there is a performance issue that is widespread. In a bad company, this is not a problem. They like it that way. The poorly prepared managers just become part of the system, quickly learn the bad habits and bad attitudes of their peers and help continue the dysfunctional culture. It has worked for them this far, so it must be good for them. The few good people who get into the company by mistake grow frustrated with the chaos and soon push the eject button. They are gone—which is fine with everyone else. No one ever follows up with the good people who leave or the mediocre people who stay. No feedback is wanted because there is no interest in the truth.

This has been a quick tour through the typical dysfunctional company, but the necessary points have been communicated. Failure is not an accident. All the decisions, values, behaviors, and actions that have created the dysfunction were selected by design.

The perpetrators may have had a different result in mind, but they had choices and picked what they wanted. The basic truism is that bad companies do not value success, so they do not get it. Success requires having a realistic view of things, valuing and paying attention to your people, identifying and fixing things that do not work, being in touch with your business, and asking a lot of questions. Bad managers in bad companies do not want to work that hard.

The reader now knows what it takes to create and sustain a bad company. The rest of the book will focus on how to identify and combat all the bad habits and behaviors found in the dysfunctional company, and then how to build the right team with the right attitudes and practices to transform the bad company into one that is worthy of the effort.

Chapter 3

Telling the Truth

Chuck: In chapter 2 we talked about the characteristics of the dysfunctional company and what needs to be addressed to transform it into a much better company. We identified four conditions that define the bad company: 1) culture of failure, 2) ineffective leadership, 3) poor strategy and processes, and 4) no value placed on people. And we made it clear that companies determine their own fate through their values and ethics, and the decisions and actions that follow. Luck, be it bad or good, has nothing to do with it. In the chapters that follow, we will go into detail about how each of these "conditions of failure" can be addressed—but first, we need to touch on some "prework" that bad companies must undertake before they are ready to tackle these big transformations. And that has to do with how virtually everyone in the bad company interacts with each other. While the people in good companies are focused on truth, reality, candid conversations, and trust, in bad companies, they do a lot of lying. And that is the crux of much of the problem.

Now, it can be argued that employee interaction is a company culture issue, and that would be correct. But it is also the key prerequisite to getting any kind of company transformation off the ground, even changing the company culture. All change processes demand that the participants are candid with each other, speak the truth, deal in reality, and have trust in each other to behave correctly and do their part of the work constructively. So Job One is making sure this happens because bad companies do not naturally operate in this manner. In this chapter, we will talk about how and why bad companies and their employees tend to avoid the truth, and what it takes to address these bad habits and show people that the way of truthfulness is the first important step on the path to becoming the member of a good company.

Why Bad Companies Lie

So far in this book, we have learned that bad companies and their people do not like reality, accountability, or any kind of oversight that involves rules or processes. They want to be left alone to do whatever they feel like doing at the time. And they certainly do not like to be questioned or corrected. They do not like facts or data and prefer to keep everything murky so anyone paying attention cannot figure out what is really happening. When people feel this way, they will do anything to keep the facts to themselves and make sure that everyone else is in the dark, including lying. Now, *lying* is a pretty strong word. We have given presentations on this topic to a room full of managers, and the initial response is usually, "That is not me. I do not lie." But then we go into what constitutes a lie and things start to change. The following are some activities in a business environment that constitute lying:

- When you mislead a coworker, supervisor, customer, or vendor

- When you hoard or withhold information from someone you know needs to be informed

- When you report authoritatively on something that you are ignorant about

- When you avoid telling bad news

- When you say you have expertise when you do not

- When you oversell a situation to make yourself look good

- When you take credit for something that someone else did

- When you destroy, hide, or change documentation to cover something up

- When you knowingly give incorrect information or data

- When you deflect or avoid a subject

- When you tell a half-truth that has some element of truth in it but is deceptive

- When you exaggerate, basic aspects are true but only to a certain degree.

This list does not cover the entire subject, but it clearly shows that when you, through your actions or through what you say, give someone an impression of things that are not totally true, then you are lying and doing all the damage that liars generally do. When those people who say they do not lie see the preceding list, they generally have to concede that they are guilty. So the big question is: Why? Isn't telling the truth easier? There are a million lies but only one truth. Eventually, lies run their course and there is an accounting. Isn't that a deterrent? Well, not really, especially in the bad company. There are a lot of reasons why people routinely lie in the business environment. We will give you five of those reasons in the following:

1. Lying is often the easier route, at least short-term. Bad companies and bad people like "easier." They just want to "kick the can down the road" or get through the meeting. They will figure something else out later.

2. Lying is everywhere and pretty much accepted practice in politics, advertising, media, and corporate communications. Quite often it is called "putting a spin on things," but it is still about separating people from the truth.

3. CEOs and other company leaders set the standard. We have never seen a bad company that did not have a liar at the top. Bad companies have a lot of bad news to report. It is often easier to create a new reality and lie about all of it. It becomes SOP within the company. If you do not lie, then you are an outsider and a troublemaker. People who lie "get it." Everyone else is just naïve.

4. People tend to rationalize the bad things they say and do. A policeman one of us knows said, "I have never arrested anyone who thought they were guilty." People say, "It was the best of a lot of bad choices." The good choices were there, but they were tough and were not considered.

5. The truth, and the reality it brings, can be hard and it can be painful. There can be a lot of difficult conversations when you stick with the truth, and employees can become angry and leave. The long term will be better, but many people do not place much value on the long term. Bad companies and their people do not like any of this.

Before we address how to get on the right path and stay there, let's go over some examples of people who did not get on the right path. We believe using real-life examples in a business environment helps readers see how this all applies to them. None of this is theoretical. We have seen these generic examples many times over.

Example No. 1

You are participating in a cross-functional team project to build some new capability in the company. You do not like the other people on the team. They are not from your group. You have some critical information that the team needs to be successful. You either deny that you have it or only divulge a piece that, by itself, is actually misleading. You let the project fail and then partially rescue it after the damage is done.

Example No. 2

You oversell a special project to management by saying you have done all the necessary groundwork when you have not. The project is now in trouble and is affecting your bottom line. There are people at corporate who could help you get things back on track, but that would require that you first "come clean" about events. You opt to buy time by moving money around on the books and hope things get better next quarter.

Example No. 3

You are the manager of an industrial operation and have had an inordinate number of safety-related close calls lately. You know things are not right with your safety program implementation, but the guilty party, one of your supervisors, is a friend. Doing a deep dive on safety right now would be a lot of work and would put pressure on your friend. You decide to stay quiet and hope things work out.

Example No. 4

You are an operating manager and you have an employee who is failing. Virtually everyone knows this person is a problem and it is causing morale issues. You know you should review this person and put him on a plan, along with probation, but that requires more work and uncomfortable conversations that could get heated, and the person could leave. That just produces more work. You tell the employee that he is doing fine and hope things get better.

Example No. 5

You are a senior manager who is putting together an acquisition that you know is a disaster. However, the completed acquisition will make you look good and get you a bonus. You really had to "cook" the pro forma to get it approved. Your financial person tells you the acquisition is a big money loser. You reply, "I don't care. By the time it blows up, I will be long gone."

It is easy to see when you have a whole cadre of managers and employees spending all their time creating the havoc that we have seen in these five examples, then it is impossible for the bad company to do much more than keep the absolute basics moving. And sometimes that is not even possible. So what are these people thinking? Based on our observations, it really boils down to four basic behaviors:

1. People who do these bad things do not monitor themselves. They are too busy trying to get what they want. In fact, when you have a whole company of people totally obsessed with getting what they personally want, the company ends up looking more like a free-for-all than a business. The whole concept of teamwork or doing what is best for the company is out the window.

2. People doing bad things do not think about consequences for themselves and definitely not for others. When people are not guided by morals or consequences, then anything is possible, and that is usually what you get.

3. People doing bad things think they are invisible. No one is watching. One of the authors once asked a colleague who was a "reformed bad boy" what he was thinking when he did all those unacceptable things in his business and personal life in the past. The person got a very serious and reflective look on his face and then broke out in a huge grin and said, "I thought I was invisible." He saw himself as just smarter, quicker, and more capable than anyone else and just could not be caught. Unfortunately, like all the lies he was telling, it was just a delusion.

4. People doing bad things never ask, "What am I doing? Is there a better way?" When you are "hooked" on doing it

your way, which is the wrong way, then you have slammed the door on any opportunity to make a good decision that could save you. People end up being defined by the decisions they make in their lives. If you adopt a dysfunctional way of conducting yourself, then that becomes the only real decision you ever have to make. Everything that follows is just habit.

At this point, you—the reader—should have a pretty clear view of the damage that lying can do to a company and its people, and you understand how people lie and why they do it. The next two sections will focus on what it takes to reverse that behavior and give people a set of habits that will better serve them in their careers and in their lives.

How to Get on the Right Path (the CEO role)

This section is focused on what the CEO must do to activate this truthfulness transformation. But before we get started on how the CEO can communicate and lead this journey involving the correct way to interact with people and how to build the habits that support all of that, it would be good to start this section out with another example that sets the tone for everything that follows. Jack Welch was being interviewed toward the end of his tenure as CEO at General Electric. (We know that recently the GE star has fallen, but during his time at the helm, GE did very well, and much of what he said made sense back then and we believe it still does now. That is why we believe this example is a good one.) The interviewer questioned his policy to annually terminate the bottom 10 percent of his managers. What follows is a paraphrase of Welch's reply when the interviewer asked, "Isn't that cruel?" Welch responded, No. To have someone failing and not tell them is cruel. By telling the person the truth they can deal with the reality and move on to a company where they can perhaps do better. The main thing I owe my people is to tell them the truth. The power of this example is that it is very simple but also very specific. Too bad all CEOs do not understand, let alone embrace, this very basic but very important concept.

Bad companies and their leadership are not focused on telling their people the truth. So when a new CEO wants to turn around a bad company, that is where it all starts. The CEO can talk about his vision for the company and the reasons why the changes are needed, but the first message has to do with how the company's leaders and employees are going to interact with each other going forward. And just like Jack Welch's message, it must be very simple but to the point. There can be no room for gray area or misinterpretation here. The CEO must go over the old behaviors and make it clear that those days are gone. This may be the first strong message this CEO has given to the company so people, especially the "Rebels" and "Terrorists" that you will meet later, will be paying attention to get the measure of this new CEO. The focus here is not about winning friends but about defining values and setting standards and, yes, explaining the rules. This is the first "stake in the ground" for the overall transformation, and people need to hear a clear message about who is in charge and what is expected. The audience must hear over and over again words such as: *reality*, *candidness*, *truthfulness*, and *trust*—and why these values are important. The CEO should go on to describe how the people in the company will conduct themselves. Down the road, there will be specific training on how this is all done, but at this point, the audience must clearly understand that: 1) this is the CEO's idea, 2) he will be involved and this new direction is not going away, and 3) there is a "zero tolerance" policy for noncompliance.

Companies often try to change behaviors the diplomatic way by suggesting new things and then trying to convince people that they should do them. Now as you will see in later chapters, we believe there is an important place for engagement, and we strongly feel that transformational initiatives should not be edicts but need to involve the rank and file in their development and implementation. But this is not the place. We are talking about behavior and we are talking about what the company culture will be going forward. This is not a negotiation because that will just water everything down and nothing will ever change. The CEO's vision of how the company will look post-transformation is at stake here, and the message and the expectation must be very specific and very easy to understand. "This is what I expect. If you want to be on this team, you will do it this way. I will be watching." Anything less will not get the job done.

Once the CEO has set the expectation and then the actual work to move the transformation forward gets started, the battle is not over. People may cooperate, but they will be watching, and a number of them will be probing and testing. At this point, the CEO role is more important than ever. There will be surprises and issues that are complicated, and there will be noncompliance. The CEO must consistently stay true to the original message and not get distracted or delegate the ownership of this transformation to anyone else. People are watching, and any sign of weakness can open the floodgates. Issues should be dealt with promptly and consistently. No one gets special treatment. The CEO should stay on message and bring the subject up at every opportunity. This is a lot of work, but what is important to remember is that if the company cannot get over this hurdle, then any other possible improvements are doomed. Failure is not an option. That is why the CEO role is so important.

How to Get on the Right Path (the Employee Role)

Whether you are a manager or rank-and-file employee, when you hear the CEO's message, you have to take stock of how you operate and then make the necessary changes. That all sounds simple. It isn't. As we have seen, people in bad companies do not like to take stock and do not like to change. Now, we are not implying that all people in a bad company are the same. A later chapter will go into detail on the various groups of people we routinely encounter in the bad company. But for the case of simplicity, we can say that there are people in bad companies who like the way they act and see no reason to change, and there are those who may have some bad habits and have definitely become used to the bad behavior around them but are potentially salvageable. This section is for this latter group or for people who manage them.

The people in the former group need to leave the company and definitely will not get much benefit from reading this book.

Bad behavior is basically a habit. Telling lies in all their various forms is a habit. People either walk into the company with this habit or they learn it through observation and peer pressure. Like any habit, if you want to change it, then you have to change your behavior patterns. We learned in the first section of this chapter that *liars* are totally focused on what they want, do not consider consequences when making decisions, think they are invisible, and are *hooked* on doing it their way. If that is you and you want to change, then this lethal pattern of behavior needs to change. You need to embrace a new way of reacting to things, making decisions, and dealing with people. It is a process and it can be learned, but it takes practice and it takes commitment. What follows is a fairly simple set of activities that will get you on the right path:

- <u>Monitor Yourself</u>—Routinely question your motives and justify your decisions and behavior to yourself. Ask yourself, "Is this the best option, and why? Am I conveniently leaving anything out? Can I easily explain my behavior to a third party?" If the explanation gets complicated, you already are back in the pit.

- <u>Look at Your Environment</u>—There are many stakeholders in the process (company, fellow employees, your boss, direct reports, customers, vendors, public). Are they being treated fairly or is it all about you? What would these third parties think if they knew all the facts? Do they have a different perspective, and have you considered any of that in your decision-making process?

- <u>Be an Example</u>—People either work for you or with you, and they are watching. Be aware of that. You either set the standard or have an impact on it. What you do, others will copy. When you see bad behavior, address it and make it teachable. Any bad people or bad behavior that you ignore will sabotage you later and become the standard. Feel you have accountability for what you see and hear and remember that every day.

- <u>You Are Part of a Team</u>—You work for the company (and they pay you). You owe management your best efforts and best behavior. You are part of a team. Remember that. Bad people see the company as something to exploit. They are not on the team. Work together with one focus. Do your part to make that happen. Make sure the people you influence see it that way as well. Share information, mentor people, ask for help. You win when the team wins.

- <u>Do It Every Day</u>—Habits require that you focus on the right things all the time. The forces of dysfunction are strong and ever present. Let your guard down and you are doomed. Be aware of your standards and how you match up to them. Any bad behavior that you allow or ignore becomes the standard way of doing things. Act quickly, decisively, and consistently. When all is done, all you really have is your reputation and how you treated others. Guard them jealously.

These activities are not a "one and done" message. If that is the case, the company will quickly slide back into the old pattern of creating false realities and avoiding the truth. All participants from the top down have to embrace these concepts every day. Self-evaluation and self- questioning need to be a continuous part of the cultural fabric of the company going forward. Like every transformation, be it personal or company-wide, you have to win and hold the ground every day. It never gets easy, but it is worth the effort.

How do you know when a company and its people have made the quantum leap from a dysfunctional band of liars to people who embrace reality, deal in truth, and trust one another? You just have to walk the halls, sit in on some meetings, and talk to a few people. Their behavior will tell you who they are and what kind of company employs them. When you like what you see, then that company is ready to move on to the rest of this book.

SECTION II

OVERCOMING DYSFUNCTION

Chapter 4

The Change Process

Chuck: You, the reader, have just completed the first section of the book, *Manifestation of Dysfunction*. In the first two chapters, we talked about why we think the book is needed in the business world and how bad companies look and behave and the reasons for all of that. In the third chapter, we started the transformation process by stressing the importance of truth in how companies and their people interact. Now you are ready to move into the second section of the book on how dysfunctional companies can be transformed. Chapter 4 is a natural beginning because it describes in a generic sense how you can make change happen in the change-averse world of the bad company. All the chapters that follow in this section feed off this very basic, but very necessary, chapter 4.

Fixing bad companies involves a lot of change. As we have seen in the previous three chapters, the company leadership, the company culture, and a significant part of the workforce and the way those people operate have to undergo dramatic change. These are not tweaks but are massive makeovers. The chapters that follow in this book describe the CEO role in all change efforts, building the guiding coalition, changing the company culture, key initiatives, transforming safety, the role of HR in upgrading the workforce, creating a performance management capability, and building and supporting key processes. These are all necessary components of the overall transformation of the company that must be completed successfully to take the company from "bad" to "good." There are no shortcuts. Each of these components is vital and cannot be eliminated or done partially. And each of these components involves significant change for the company. What we have learned so far is that bad companies do not like change and the people in those bad companies are very good at derailing efforts to change anything in the company. So this transformation is going to be a war that the CEO and his team must win completely and definitively if the company is to emerge as something significantly better than it was originally. The good news is that for all of these transformation components, there is only one way to make the change happen. In each case, the details will be different, but the basic process is always the same. In this chapter, we will describe how this generic change process looks and how it works. In the chapters that follow, we will go into the details that apply to each of the components.

Mobilize Leadership/Identify the Need

All change efforts require a strong leader who can not only lead the change but own it. We started to talk about that in chapter 3, and it will be a recurring theme throughout the rest of the book. In many cases that strong leader is the CEO, but it could be another officer in the company. Whoever is in that role, that person must command respect due to his authority, ability, and character. CEOs often make the mistake of getting a change effort started and then handing it off to someone who is not up to the task. The people opposed to the change, whom you will meet in chapter 6, will eat this person alive and kill the transformation before it hardly gets started. The leader must be a formidable person who can stand up to strong opposition and make it all happen. "Owning the change" means just that. The leader must tell everyone up front, "This is my project. I believe in it. I am involved. I am not going away and I will not lose heart. If you oppose this project, then you will be dealing with me." Change projects die when they get passed around and become orphans. Successful change projects always have a home. The leader must be on top of everything all the time, ready to deal with issues, and do whatever it takes to keep the project moving forward in the right direction.

John Kotter, in his book *Leading Change*, says that all change efforts must get off on the right foot by defining the crisis and creating a sense of urgency. What that means is that the leader must make a strong case for the change that everyone can understand and believe. Transformation involves taking a group of people moving in one direction and getting them to change that direction. That is a huge undertaking. You cannot get that done in a sustainable way by just telling them to do it. People have to hear a persuasive argument that most can believe that defines the problem and communicates just how important the need for change really is. Just making the effort to engage and inform the workforce will go a long way toward building a connection. These people usually do not see this coming from the leaders they have had in the past. It will get their attention, and many will start to get involved in fixing the problem.

By taking these two critical steps right at the beginning, everyone now knows what the problem is, who is in charge, and at least in general terms what that leader wants to do about it. Now it is time to broaden the capability needed to get the transformation effort moving and determine what needs to be done.

Build the Team/Build the Plan

John Kotter also says that any change effort needs a guiding coalition to move the transformation forward. This guiding coalition is basically the team that builds the transformation plan and then drives the activation of that plan "out in the field." The key to building a successful team is getting the right combination of people on that team. First, the team must have the necessary hands-on leadership to represent the program leader on a daily basis, deal with difficult issues, and keep things moving in a positive way in the face of intense opposition. People of this quality are rarely found in bad companies and, if present, they usually do not have the credibility they deserve. In most cases, these people will have to be brought in from the outside. Many change projects fail because coalition leaders are not up to the necessary standard. This is not a place to compromise on quality just to get positions filled. These people have to be impressive and know how to exercise effective leadership in all kinds of difficult situations. The second category of team members are the "knowledge experts," who bring the technical capabilities needed for whatever the change effort is intended to address. For example, a safety change process would require team members who are experienced in building first-rate safety programs and understand all the program components that are required. Most of these people will also have to be brought into the company. However, if there are a few incumbent employees who have some of the needed capabilities, it can be a positive to include them, as they understand what is there now and how the various existing managers will respond to the changes. The last group of people is often left off the team, and this is a critical mistake. This group consists of incumbent managers and employees. You may ask, "Why do we need these people on the team?" They are important for several reasons. First, without them, the team is basically a bunch of outsiders with no connection to the company. This is bad optics when you are trying to engage a workforce in a large change effort. Second, these incumbent people do understand the company and can provide perspectives that are needed and can be surprisingly informative. Third, the team needs to build a connection point within the company, and it helps when team members are part of the target audience for the transformation. Needless to say, selecting incumbents to be on the team can be tricky. You need people who are excited about the change and want to contribute as well as people who have reservations but are open to being convinced if the plan makes sense. There should be some manager-level people on the team, as they can help move the program forward during implementation and many people see them as having credibility or at least local authority. Last, it can be useful to have a few upper-level people who are opposed to the program—provided their actions are not obstructive. These people can raise important issues that could go overlooked but must be addressed. Having them close at hand gives the program leaders ample opportunity to understand their issues and ultimately get them involved in the change and actually supporting it. Teams are hard to build and sometimes mistakes are made or people do not behave as expected. Team leadership needs to be on the lookout for problems and be ready to responsively deal with issues to keep the program moving forward. In the final analysis, this team has to build the plan and then make it happen in the field, so they have to be up to that effort and committed to making it all work.

Building the plan is much like building the team. The rank and file must ultimately accept this plan and get comfortable with it, so how it is built and how it is viewed are just as important as what it contains. The plan will include some generic components that come from the outside and are necessary ingredients for any change effort involving the topic being addressed. However, in addition, there must be custom-built components that specifically address issues in this company and are critical to making the program successful for this application. A good leader will use all the capabilities of the coalition to build the plan that is needed. He will listen to all the viewpoints and try to build broad support across the coalition for what is being developed, but at the same time not sacrifice personal standards for the quality of the final product. It is also important for the leader to keep in mind the overall capability of the company as a whole and not to "boil the ocean" with a plan that is too big or too complicated. Furthermore, it is vital to keep the process moving. Bad companies have short attention spans and the negative forces will try to bog the effort down in the hopes that it will lose steam and people will walk away from the effort. The leader and his key associates must keep the program fresh and in people's minds by making legitimate progress and keeping the rank and file apprised of that progress. At the end of the day, the leader must be proud of the final product, confident that it will work, and feeling that the coalition is behind the plan and ready to implement it and support it.

It can happen that key issues pop up right at the end of the plan-development phase. At times, these issues are a ruse, perpetrated by the opposing forces to delay everything, but sometimes these issues are legitimate or at least much of the coalition thinks they are legitimate. The bad leader will shut everyone up and trample over the group in order to get the plan out the door and into the implementation phase. This is a massive mistake. After taking great pains to involve a broad spectrum of the company and to listen to all the perspectives, the bad leader has destroyed all the credibility that has been built and shown his true colors by shutting everyone down and doing what he wants to do. Good luck integrating that plan into the company. The good leader will move fast, but he will listen to the concerns and make the necessary adjustments in the right way and stay true to the process that was originally adopted to build the plan. The good leader knows that when that plan goes out to the company, he needs all the good-will he has developed so far in the change effort. The program for change is not almost done at this point but really only in the third inning. The hard work of getting people to accept the changes, learn the processes, and then support them long-term is the real battlefield for this effort. And that comes next.

Setting Expectations/Following Through

We have talked about how important it is to stay in touch with the rank and file and keep them in the loop about what is happening. This is another key time to do just that. Preferably the CEO or at least the program leader must communicate what he expects will happen next. This is a continuation of the initial statement the leader made when the problem was defined and the commitment was made to fix it. Now is the time to lay out for everyone what the plan is and what it will accomplish. This is a call to action. The leader must reiterate that he is committed to this change, will be directly involved, and will see it through to completion. He should state that he has been directly involved in building the team and building the plan. The team is up to the task and has taken great pains to get all the perspectives and build a plan that can do the job. The plan meets the leader's standards and the leader is enthusiastic about sharing it with the rank and file and working with them to use the plan to improve the company. People want to know *what is in it for them*, so this is an ideal time to explain how the changes will improve the company and how those changes will make people more efficient and effective. The goal here is to get everyone on the same page and make sure no one feels left out. If everything is done correctly to this point, the company as a whole will know what to expect, people will begin to understand their roles, and everyone will have a clear idea of what will be gained by the effort. And many of the people will begin to get excited about what is going to happen.

We referenced the book *Execution* back in chapter 1 and we will reference it again several times in this book. In that book, it states that leaders must *follow through* if they want to see their plans become a reality. We could have used the term *program activation* or something like it to define this section, but *follow through* really strikes at the heart of what must be done next. In our minds, this term means turning words into actions. Many leaders believe they are only responsible for the words; other people are responsible for the actions. That is a recipe for failure. In this particular case, the bad leader would make the speeches and then turn everything over to someone else. People would immediately see that the leader was not serious and they would begin to test the new guy. The good leader makes it plain up front that he will be on the front lines. He is not a micromanager but will be involved. The good leader wants to see firsthand how the various program modules are implemented and wants to get direct feedback on how it is all going. If there are issues, he wants to know. The good leader is involved, is responsible, and is accountable. That is a new concept for bad company employees. They are used to everyone distancing themselves from outcomes. Here is a leader who is grabbing hold of an important transformation project and announcing to everyone that he will lead this effort to success. That sets the tone for everything that must follow.

Like the program leader, coalition members must be on the front lines working directly with people to help them understand the plan and how it works for them. There may be professional trainers and subject matter experts involved, but coalition members are vital to the effort. The team needs to know, in real time, how the plan is working and how people are accepting it. If there are problems with the plan, they must be addressed and fixed quickly. Driving a bad plan down people's throats is a great way to destroy a change project. It is critical that people have a chance to talk, ask questions, and voice their opinions. They will only come around and support the plan if it works and they know that the team is focused on making it work for them. This is also a great way to get the measure of people, especially managers, out in the field. Some people will step up as leaders for the effort, offer great ideas for improvements, and become resources for implementation. These people may end up being one of the long-term owners of the new program being implemented. Other people will push back and show no interest in working with the team, even after their concerns have been addressed. These people need to go, and this is a great time to do just that. In fact, not doing something about these people sends the message that bad behavior is acceptable. People need to understand that they will be listened to and they will be helped, but they must cooperate and be part of the team and part of the transformation.

Expand Ownership/ Institutionalize the Change

People often think that once they build the plan and train people on its various modules, they are done. That is far from the truth. As the old saying goes, "you cannot push a string," which in this case means you cannot make a company embrace a new idea by just pushing hard from the top. The good leader knows that he must build belief and ownership within the ranks of the employee base. In successful transformations, the coalition works hard to identify and recruit people at the manager and employee level who demonstrate high levels of interest, capability, and leadership during training and installation of the various modules. People are more prone to embrace new ideas that come from leaders and peers that they know and respect. Having a cadre of incumbent employees who are "on the team" is the only way to ingrain a new idea or a new process deep into the organization and have it become a natural part of the way the company thinks and operates. As this piece of the transformation process matures, a part of the ownership of the change naturally moves from the coalition to these local process leaders embedded in the organization. They take on much of the responsibility for supporting the process and ultimately finding ways to improve it. The smart leader will cultivate and reward these "local process leaders" with acknowledgment as well as future opportunities for advancement. Not only is this a great way to get the most out of the best people in the organization; it is a strong message to everyone about the many benefits of getting on the team and helping the change happen.

Even with a large network of local process leaders in place and the change apparently firmly embedded in the organization, the transformation leader and his coalition cannot rest and are far from done. The opposing forces never give up as long as they are present within the company, and they will be fighting hard to make this change temporary. They will put pressure on every manager and employee they touch not to use the change, or at least to marginalize it. The new local process leaders will come under tremendous pressure from people who can affect their careers to back off and be less of an advocate for the change. The coalition has to keep working the system, one manager at a time, to make sure things are running smoothly and there is no terrorist activity within the ranks. People who are using the change and are advocating the change need continuous support from the top. Asking them to stand alone is asking too much and is putting the transformation at risk. The transformation leader (and the CEO if that person is not the transformation leader) needs to keep the topic of the transformation fresh by talking about it and the progress that has been made or the challenges that must be overcome at every opportunity. People need to know that company leadership is watching, is on point, and still cares about the change and how the rank and file is dealing with that change. If bad behavior is identified, it must be dealt with quickly and in a very consistent and decisive manner. Managers and the employee base need to see that there is no room in the organization for people who want to sabotage the transformation. It makes no difference who they are, the position they hold, or how long they have been around. If they are not on board, they need to go. Leaders make the mistake of tolerating bad actors in high positions. That is a fatal mistake. People see that it is okay to push back and then the transformation is lost. If the change leader and the coalition stay true to these principles, the change will ultimately become a natural way of thinking and doing within the company. When all managers are using the change and expecting their people to use the change, and when new people are trained that "this is the way we do things," then the change has been institutionalized.

Celebrate Success/ Keep It Going

We have talked previously about how the natural forces of chaos are very strong, especially in the bad company, and that organization against chaos requires not only massive but continuous effort. That is very true at this stage in the transformation process. Although things have gone well up to this point, the whole project can wither away if leadership loses focus, thinks they are done, and gets distracted by other issues. What must happen next is not just cleaning up but constitutes a vital part of the whole effort. If you miss this, you can lose everything.

First, this is no time for complacency over success and a show of false humility. This successful attempt at change is a big deal for the bad company. They have never done this before, and for most of their careers the managers and employee base in the company have believed that they were not up to any of this. They are not used to winning. They have only known how to survive. So the transformation leader and the coalition need to celebrate this success by acknowledging that the whole company has *grown into this* and that they all are better for it. This is not a "one and done effort" but a new beginning. Who they are and how they think and act has changed. The leadership should take this opportunity to state that, by working together and using the right principles, they have demonstrated that they can do great things. This is the first big step up the ladder to becoming a much better company. The leader is proud of what has been accomplished and is looking forward to doing more of it. That sets a new expectation and keeps the energy and the confidence across the company at a high level. That original sense of urgency must be sustained.

A company that was about three years into a major safety initiative that was doing well was having a senior-level planning meeting. The topic of safety came up and a team member said, "Well, we are pretty much done with that one." Most everyone agreed. At that point, the future of that particular initiative was doomed. Complacency had begun to set in and everyone was moving on. Safety, at least at the top of that company, was becoming an orphan. People at the lower levels in the company could try to keep it going, but ultimately the natural forces of chaos would win. That is why "keeping it going" is crucial to any change. Now what exactly does that mean? Things stop being active and meaningful when the people at all levels no longer talk about them, they disappear from employee goals, and leaders quit asking questions about them. So it follows that "keeping it going" means keeping the topic alive. And that process starts from the top down. The CEO and/or transformation leader should include the topic in every major communication to the company, and in his goals and reviews for direct reports. When senior leaders visit local operations, those managers should be asked about how the new change is going and what they are doing to make it more effective. Local process leaders should have face time with senior leaders so they can give direct feedback on what they are seeing, where they need help, and how the new process can be improved. If senior leadership can consistently demonstrate through their actions what they want done, the next level of management will figure it out and the behavior will trickle down through the organization. When people show complacency about a "completed" transformation module, they need to be confronted and corrected on the spot. Too many leaders tolerate these destructive attitudes and behaviors, and then they wonder why things are not going as well as they would like them to be. Strong leaders are not shy about communicating what they want done and then showing by example how to do it. Setting the standard and then living that standard is a big part of "keeping it going."

Now you have seen how to get a change process started, how to engage the company and get them involved, and how to institutionalize that change into the very fabric of the company. What is very apparent is that making such a change happen is a lot of work and requires great courage and commitment. Leaders often think they can create a great plan on their own and then the employee base will just accept it because it is so brilliant. Any accomplished marketer will tell you that products and ideas do not sell themselves. Good leaders know that the plan has to be great, but the effort to engage the company and get them believing in the plan and in themselves has to be even greater. That takes a lot of face time with a lot of people, in which leaders listen as much as they talk. This message will be reemphasized over and over again as we make our way through the rest of this book and see how the change process is applied to each particular application.

Chapter 5

Fixing Dysfunction

Jim: We have discussed in detail some ingredients of dysfunction. In this chapter, and those that follow, we'll offer a recipe that helps managers recognize, diagnose, and overcome dysfunction and, in turn, transform organizational performance. It may be helpful at this point to recap what we mean by dysfunction in the context of this book. As defined by Merriam-Webster, dysfunction is abnormal or unhealthy interpersonal behavior or interaction within a group. As we outlined in chapter 1, organizational dysfunction, for our purposes here, is characterized by major cultural and systemic issues, such as weak, overwhelmed, or misdirected leadership. There are huge silos between groups, processes are ineffective or nonexistent, and accountability is absent. People are poorly trained, poorly managed, and not motivated to succeed. There is an underlying culture of failure where management and employees avoid reality and operate within the illusion that everything is fine. People at all levels are in survival mode, there is little or no employee trust in leadership, and their fear of making a mistake or alienating a person of status is more important than accomplishing something. In sum, there is no effective strategy for success and, in fact, success is not even valued by the organization.

The Basic Realities

There are six key tenets of dysfunction that must be understood in order to turn negative inertia into positive energy in any organization.

1. Dysfunctional routines will develop in most any organization, irrespective of group size or charter.

2. Dysfunctions can range in scope, from obvious to obscure.

3. Organizational dysfunction will not self-repair.

4. Healing dysfunction is not a grass-roots effort—it must be driven from the top.

5. Stay the course. Expect headwinds and rough skies for a while.

6. Left unattended, organizational dysfunction will marginalize productivity, efficiency, and bottom-line results.

Let's look at each in a little more detail. As stated, it is critical to understand and expect the behaviors in order to lead transformational change in any engagement.

1. Dysfunctional routines will develop in most any organization, irrespective of group size or charter.

   In our experience, it is inevitable that some degree of dysfunction will appear sooner or later when two or more people interact on an endeavor. Some call it *organizational cholesterol*. Others call it *bureaucracy creep*. The more people there are in the mix, the more opportunity there is for a culture of failure to develop.

2. Dysfunctions can range in scope, from obvious to obscure.

We have been part of organizations with tens of thousands of employees, and we have been part of organizations with fewer than twenty employees. We have served in both private and public sectors, and we have worked internationally. While there is no template or matrix that prescribes the kind of dysfunction that will develop in companies, the remarkable lesson we learned is that it is a very safe bet that dysfunctions will manifest in some fashion—be it glaring and overt or be it subtle and hidden deep in the social fabric of the group.

3. Organizational dysfunction will not self-repair.

   Identifying and addressing destructive routines takes a deliberate engagement by management. These routines will not evolve and self-fix. The longer it takes management to take on the problems, the deeper the cultural scar.

4. Healing dysfunction is not a grass-roots effort—it must be driven from the top.

   Some managers believe that, given enough time, dysfunctional practices will self-heal if just left alone. That somehow things will work out eventually. That people in the trenches will naturally overcome and the job will get done. Wrong. Left unattended, organizational dysfunction will marginalize productivity, efficiency, and bottom-line results.

5. Stay the course. Expect headwinds and rough skies for a while.

   Like anything that offers a big payout, this can be hard work. It can be uncomfortable to challenge organizational dysfunction that may have taken root and is a long-standing cultural custom. Sweat is a great solvent, and it quickly washes away tepid leadership. Leadership, patience, and commitment will be tested. For those leaders who do stay hard on task, despite the temporary discomfort, the rewards are waiting. For example, a number of years ago an innovative CEO in a major transportation company recognized that his management team's poor relationship with the labor union leadership and employees dramatically lowered the value of the franchise and inhibited his ability to provide shareholders a maximum return on their investment. It was also causing turmoil, safety problems, and customer dissatisfaction. The labor unions represented the vast majority of the company's frontline workforce, and decades of mistrust and acrimony had resulted in lines being drawn between management and the unions. Not a healthy environment to take on a major acquisition that was pending. This CEO brought on board a high-energy change agent with wide credibility in both management and labor circles to bridge the gaps and get people back to the table. The naysayers and established management in the company covertly pushed back, dragging their feet, and quickly declared failure at the first hiccup. It didn't deter this CEO. He remained patient, steadfast, and focused. Eventually, people started to see the benefit of working together to resolve myriad issues around the major acquisition under way. Bottom line: the acquisition was successfully accomplished and the company ran past its closest competitor.

6. Left unattended, organizational dysfunction will marginalize productivity, efficiency, and bottom-line results.

Dysfunctional routines consume time, energy, and resources. Left unaddressed, dysfunction will undermine organizational effectiveness and result in the squandering of physical, financial, intellectual, and emotional resources. Simply put, dysfunction gets in the way of people trying to do their jobs. Dysfunctional routines demand time and attention that could be more profitably applied to meaningful business issues that benefit all stakeholders.

Diagnosing Dysfunctional Routines

Now that we understand the six basic underlying precepts of organizational dysfunction, let's turn to some thoughts around the diagnosis of dysfunction.

First, it is important to note that you can't effectively identify and diagnose dysfunctions sitting behind a desk in the executive suite. In our experience, many senior-level managers believe that they are in touch with reality simply by virtue of their position. As discussed in the next chapter, the CEO can be especially vulnerable. We call it the "CEO disease"—where a CEO believes he is right about everything. Some CEOs and their minions tend to create their own reality, which is likely not the organizational reality at all. They order repairs based upon their flawed view of reality. It is classic management malpractice where remedies are prescribed without first validating the diagnosis.

The only surefire way to understand the cultural goings-on in the trenches is to see them firsthand. To touch them, smell them, and watch them. To regularly "cross-check your gauges." The Honorable Robert L. Sumwalt III, National Transportation Safety Board (NTSB), and a retired airline pilot and major corporate flight department manager, talks about the importance of aircraft engines to a pilot flying an airplane. The engines propel the craft, overcoming drag. They are essential to flight. Sumwalt says that given this critical function, we doubt anyone would be satisfied with a single diagnostic indication on a cockpit instrument to reflect the health of the engines. For example, a lone gauge that simply displays "engine okay" or "engine not okay" doesn't provide the kind of information required by the pilot. Instead, engines on airplanes have multiple sensors imbedded throughout their systems to monitor and report critical engine health and functionality to the pilots through a series of cockpit instruments.

Sumwalt makes the analogy that the engine that propels companies is the culture. How things get done. "As a senior manager," he asks, "wouldn't you want to know how well your 'engine' is functioning? Like a pilot who has benefit of many sensors monitoring aircraft engine performance, senior managers also have 'sensors' placed throughout their organizations." Those sensors are the employees, the frontline supervisors, contractors, and customers. Managers who recognize and solicit input from the *sensors* in their organizations are most effective in identifying and addressing creeping dysfunction before gaps develop and damage ensues.

Said another way, listen to the drumbeat coming from the ranks. Your organization will talk to you if you are willing to hear. Long before a dysfunction impairs governance or diminishes financial or operational results, there are signals. It might be poor safety performance, customer complaints and defections, budget deviations, silos between departments such as marketing/sales and the operating function, labor discontent, or high supervisory turnover. If you listen with awareness, you will hear the drumbeat before problems take root.

There is a caution we offer here. Some senior managers get "target fixation" and that leads to loss of enterprise awareness. Target fixation, to a combat pilot, is the tendency to focus exclusively on the target in front of you. When this happens, there is a loss of perspective and situational awareness about the totality of your environment and what else is going on around you. In combat flying, it can kill you. In business, it can be a going-out-of-business threat by marginalizing franchise potential with imprudent allocation of resources and capital dollars.

Let's say there was a CEO of a company who focused almost exclusively on cost-cutting strategies with little interest in anything else. Initially, the strategies helped boost revenues and shareholders were pleased. As time went on, however, high-margin customer defections increased as services and reliability thinned with reduced capabilities from over focus on budget instead of customer value add. Since the CEO and the senior staff had target fixation on cost cutting, there was reduced attention on operational execution. As service deteriorated, customers left. The *CEO disease* blinded the top person to the reality, despite pleas from field management and line sales representatives. He knew more than anyone ....or so he thought. The outcome is that the company underperformed, many frontline supervisors disengaged, they lost some high-margin customers, the stock languished, and the brand suffered in the public domain.

Companies that are mired in dysfunction are often absorbed in a lifestyle of "happy talk." Candid conversations are not encouraged, nor are they sanctioned. In fact, anyone who attempts candid conversation is considered a dissident. This further deepens the dysfunction. This was most apparent early in the rebuild of a company's safety culture after abysmal casualty losses. A new safety VP was recruited from the outside to bring a fresh perspective and question status quo. Candid conversations were required. However, the first time he attempted to question program implementation with a long-tenured incumbent field director, he was challenged by the man's senior manager for being too direct— "hurting the feelings" of the field director.

Our experience has shown that dysfunctions generally inflict most harm during the critical time between strategy definition and strategy execution in companies. As we discussed in chapter 2, there are four main organizational attributes that feed dysfunction:

- A culture of failure

- Ineffective leadership

- Poor strategy and processes

- No value placed on people

Diagnosing dysfunction doesn't require capital, additional headcount, or a consultant. It requires engaged, attentive leadership paying attention to the basic governance, listening to the organization, and regularly "cross-checking the gauges."

Some Common Organizational Dysfunctions

In our experience, we have encountered three main kinds of dysfunction that plague organizations:

- Organizationally generated

- Individually generated

- Externally generated

Organizationally Generated Dysfunction

<u>Falling out of formation</u>—To maximize financial and operational results, everyone in the organization must be aligned. They must have full understanding of the organizational vision, values, and mission. In a March 2011 interview published in *McKinsey Quarterly*, Bombardier CEO Pierre Beaudoin said, "Our employees said it was very hard for them to support where the company was going because they didn't know what we really valued as an organization. In fact, we'd asked our employees what objectives they thought we valued, and although we had very big strategic plans, nobody could answer the question." For any team to pull in the same direction, it has to know what you're looking for and feel a connection.

<u>Got chaos?</u>—Some people thrive when things are chaotic. It becomes their mission to create a flurry of meaningless disorder. This enables the masking of accountability and nondelivery of results by creating turmoil. In this routine everybody has to approve, but it only takes one "No" to cut off creativity. Cynics find good reason not to make change or engage. "Too expensive," "We have never had an accident," "You don't understand our operation," etc. In this scenario, there is delight in spoiling creativity and initiative.

The grass is always greener—Rather than face the problems of the culture, here people try to deflect attention with new programs that do little but mask the foundational problems. Great emphasis is placed on process, not content. There is much activity around what is in the *showroom window* but little in the *stockroom* to back it up. Here, the form of the presentation trumps substance or ability. There is very little headway ever made and little is ever accomplished.

This too shall pass (or the ever-popular *flavor of the month*)— Senior management fails to stick with initiatives after much fanfare in rollout. That leads to organizational cynicism for any new program. A bunker mentality takes root in the ranks. People learn to keep heads down since this too shall pass. It is an organizational death spiral.

Fire, ready, aim—This is where management reacts, takes action, and makes commitments before understanding the root cause of an issue. We talked about it earlier in this chapter. Someone once said that prescription without diagnosis is management malpractice. Or on the other end of the spectrum, the organization is unable to reach a decision and take a definitive course of action. There is continual tinkering and fine-tuning, but nobody ever pulls the trigger.

In this corner—This is where management pits departments and groups against each other for limited resources or attention. It results in deep silos and unhealthy conflict. There is no alignment and no team effort. There is no appreciation for the art of collaboration—people believe there have to be winners and losers—not what is best for the customer and shareholders.

There are no rearview mirrors on jet airliners—This is a lesson we learned from Gordon Bethune, former CEO of Continental Airlines. His point—what happens behind a jet airliner doesn't matter at 500 mph. As organizations move forward and plan for the future, some people want to look in the rearview mirror and repeatedly hit the replay button. Whether it happened last week or five years ago doesn't matter. These people constantly look back and play "What if," adding little value to future achievement.

<u>Politics isn't pretty</u>—In many organizations, people waste a lot of time and energy trying to play politics instead of fixing problems and maximizing results for the end users. One example that was particularly troublesome involved the interaction of the two top executives at a big company. Both had been competing to replace the incumbent CEO, who was retiring. When one of them was named the new CEO, the other was given the consolation prize: second in command. Quickly the company divided into two respective camps, depending upon where you ended up after the dust settled. Many good people in the company just wanted to do their jobs and focus on job execution, but they found themselves unwillingly yanked into the political dance. Candid conversations were out for fear that something said would result in disfavor by those in one camp or the other. This evolved into a lot of wasted energy, make-work, and silly decisions.

<u>Suffering the soapboxer</u>—Most of us have encountered a coworker who feels compelled to orate on every issue, convinced that his view is the only right view. Like some street-corner evangelists, they talk at you with self-righteous arrogance. In the workplace, the soapboxer will try to dominate meetings and hijack the agenda. In her article *How to Deal with An Office Soapboxer* (*Harvard Business Review*, August 30, 2016), Alicia Bassuk wrote: "At work, a soapboxer tends to be utterly convinced that his or her view is the *only* view—and vocalizes it. Being near such a person can be unpleasant, annoying, and antagonizing. Trying to work with one, especially during a group task, can be alienating and incredibly unproductive. A soapboxer can elevate tensions to point of completely destroying the rapport of a well-functioning group." Left unchallenged, the soapboxer will eat away at team morale, promote tension, and discourage your hard workers.

Individually Generated Dysfunction

It isn't the arrow ....it's the archer—Too many people today want to blame something or someone for every failure or mistake. Like the archer who blames the arrow for missing the target. It is always something or someone else, and it is a victim mentality. Unless and until there is a culture of accountability where people at all levels believe the buck stops with them, and accept responsibility for their performance, there will be marginal outputs and mediocre company performance at best. As Amazon founder Jeff Bezos said in a 2013 TV interview, "Complaining isn't a strategy."

Father knows best—Here a dominant boss or senior manager views himself as the adult in the relationship. This is the meddling, overinvolved boss who can't help but micromanage every function in his domain. This is also the person who takes credit for all good that comes from the efforts of the people. Good ideas somehow become his ideas.

Changing a tire on a moving car—This is the opposite of *target fixation* we discussed earlier. In this scenario, CEOs and other senior managers continually change direction. They alter priorities haphazardly depending upon where scattered thoughts take them. This can destroy staff morale and add a dimension of frustration not easily overcome. As a result, productivity and output are stalled or destroyed.

And the award goes to . . .—This is where there is an overabundance of drama. It is difficult to have any meaningful collaboration with these people due to their propensity to cloak every interaction with drama and intrigue. This kind of drama can sabotage good social order and the vitality of the team is sapped.

The royalty—This is where organizational *blue bloods*[1] suffocate new ideas that challenge their sacred cows or territorial markings.

---

[1] These are people in the organization who by virtue of long tenure think that they are untouchable, exempt from the rules, and/or entitled to special treatment.

Newer employees are especially vulnerable. Left unchecked, the blue bloods will destroy initiative and good people will leave. We saw this firsthand while trying to revise the discipline policy in a large transportation company directed by the CEO. The existing policy was despised by the workforce and the unions representing them. The policy was a relic of past practice and was contrary to the establishment of a more just culture. Rather than support the initiative, the senior-most operating official in the company—a blue blood—engaged in a passive-aggressive battle with program designers and implementers. He didn't like the changes and was consumed with hallway sabotage, covertly trying to sink the program. This delayed and complicated the rollout of a very good program that served to engage and align the workforce.

Kabuki dancers—This is where people go to significant lengths to dance around an issue instead of addressing it head-on. One good example we witnessed was a major transportation company with dreadful safety performance. Instead of diagnosing the cultural problems and lack of leadership accountability, senior management threw people at it—i.e., they created more safety managers in the organization. They liked the optics, "Hey, we're adding safety people." But it had no impact on performance, because safety managers were tasked to do many of the non-safety-related chores they did before the title change (e.g., manage claims, collect tolls, handle discipline). The root cause of the problem—the lack of leadership engagement, accountability, commitment, and vision—was not addressed.

Running for election—People care more about being popular than doing what needs to be done. We saw this in a large service company. The chief operating officer was so consumed with being popular with the field management team that he risked making poor decisions that hurt the company—just to be considered a *good guy* with his constituents.

Externally Generated Dysfunction

Mischief-makers—These are people or groups from outside of the organization that impose an unreasonable workload through nuisance complaints or queries. This requires staff to pivot away from more consequential actions that directly promote productive business outcomes. For example, there was a social do-gooder who believed he was ordained to harass the staff of a federal agency charged with safety oversight of a major transportation industry. Every day, he believed that his mission was to call, write, and pester staff with questions, requests for data, and sometimes leveling frivolous allegations that could not be ignored by staff. In one instance, he accused a senior manager of destroying files and ignoring public inquiries. It was not true, but nevertheless, staff had to rally to prove it, thereby taking them away from more significant activities. These mischief-makers can cause significant turmoil and detract people from focusing on issues of real substance and efficient governance.

A bored board—In our experience, some boards of directors are incapable of offering bottom-line value to the company's mission. On the other hand, however, board members can cause turmoil and generate extra work by unnecessary meddling in day-to-day details. Some board members view themselves as the ultimate process cops and, like the mischief-makers, they impose non-mission-critical queries that consume staff time. For example, there was a board member in one of our companies who regularly nodded off during meetings when staff was updating certain operating protocols around employee certification. But this didn't stop him from sending inquiries on a regular basis asking for very detailed information about how managers were observing and evaluating the processes. He asked to be copied on routine reports, and even called field managers directly. This caused considerable disruption and aggravation at the working level and detracted frontline managers from focusing on proper execution.

Pressure cooker economics—Sometimes the economy and business cycles impose significant strain on company resources and management skills. This may result in loss of focus due to heightened pressure to meet shareholder and stakeholder expectations. The management may go into survival mode, shedding smart processes and introducing additional dysfunction into an already fragile environment. In his book *Driving to Perfection* (2012), Brian Fielkow writes that smart managers never waste a good crisis. By that, he means that smart leaders use bad times, be they economic or otherwise, to rally the troops around the company flagpole. To reaffirm the values and recharge alignment. "It is in these times of threat," writes Fielkow, "that smart management can accelerate employee loyalty and engagement and come out stronger and more culturally fit." Unfortunately, many companies do just the opposite—they shrink in times of crisis.

Some Cornerstone Principles to Overcome Dysfunction: The Human Capital

In this section, we'll focus on the human side of the equation. You can gold-plate facilities, sharpen processes and metrics, and purchase the best equipment and tools. But none of it matters if you don't offer a good product and your employees don't like coming to work. Employees who don't engage and align around organizational vision, values, and mission will easily annul any management attempt to maximize organizational performance. Gordon Bethune, former CEO of Continental Airlines, and architect of the 'worst to first' transformation of the company, was asked in a presentation to a group of investors several years ago why he talks so much about people as the critical ingredient in Continental's improvements. His answer: "Because I used to be one of them." His point is well-taken. Don't forget what it is like to be in the trenches and what was then important to you.

There are several underlying theories that are key to understanding how to repair social and/or structural dysfunction:

<u>Self-interest</u>—As suggested by Len Fisher in his book *Rock, Paper, Scissors: Game Theory in Everyday Life*, self-interest is a driving human motivation. Our experience has been that an effective leader gets it, understanding that to fully engage people and fix dysfunction one must make it personal and appeal to the basic *What is in it for me?* mentality.

<u>What people value</u>—People need more than a paycheck. Money and benefits are required to sustain families and individuals. But people value the things that money won't buy. It is respect, inclusion, feeling valued and appreciated. A variation of an adage often repeated is that people will forget what you said, they will forget what you did, but they will never forget how you made them feel. To affirm the point, we suggest a quick survey. Ask your employees to name the last three Super Bowl MVPs, or the last three Best Actress winners, or the last three Nobel Peace Prize winners. Chances are people won't be able to do it, despite the fact that these are not insignificant events or celebrities. Then ask them to name a teacher or friend who made a difference in their lives or helped them through a tough time. The lesson here is that it isn't the rich and famous who change lives. It is people who demonstrate interest in you as a person. This theory, that people value most the things they can't buy, has been affirmed in many studies, including a Right Management study of engagement involving nearly 30,000 employees in fifteen countries worldwide. In that study, rewards/compensation was well down the list of qualities employees valued.

<u>The human essentials</u>—Being human is a full-time job. You can't turn on and off human properties. That is, you can't be a human off duty and a machine or something else on duty. Good or bad, the human has several basic needs that must be met to be successful and productive at work:

- Physical needs. People need to feel like they work in an atmosphere wherein their boss considers and protects their physical well-being.

- Mental needs. The need to self-express, be creative, and be challenged.

- Emotional needs. People need to feel accepted with a sense of self-worth. That their toil is appreciated.

- Social needs. People need to feel like they are part of something special and that they work in a craft that people respect.

- Spiritual needs. This isn't a referral to matters of religiosity. Instead, it is a belief that people need to have a sense of purpose. That they are making a difference in the world.

<u>People must be heard before they can hear</u>—In dysfunctional companies, employees are talked *to*. In healthy companies, employees are talked *with*. There is a big difference. People who feel like they are first heard are much more likely to hear what management has to say. The importance of maintaining an active and healthy dialogue with employees and frontline supervisors is one of the distinguishing attributes of successful companies.

<u>Life in the shadows</u>—Henry David Thoreau wrote that most men are slaves to their work. He concluded: "The mass of men lead lives of quiet desperation." We have found that to be a truism in many organizations. There is a lot that management can do to overcome this sense of career hopelessness. Many people spend their career lives locked in the shadows, hoping that one day they will experience fleeting moments of significance. For many, time has moved on and dreams have died. They find themselves in jobs that are small and unfitting for their inner souls. This creates a downhearted worker who lacks energy and pride in the company's mission. Smart managers recognize the importance of creating an aura of significance for every job. For example, a senior-level manager on a major railroad was visiting a train yard when he encountered a worker. The manager introduced himself and asked the worker her name and what she did for the company. The worker

said, "Oh, sir, I am only a conductor." The manager replied, "Only a conductor? Do you realize how important you are to our customers? Do you realize that at any one time you are the one who is trusted with a trainload of important goods critical to a customer's success? Do you realize that you and your crew are entrusted to safely move hazardous materials through towns and neighborhoods, past schools and homes? *Just a conductor?* Without you and others like you, we don't survive as a company." It was a revelation to the conductor. She immediately saw her role in a much different light. She was energized.

Another example involved a large garbage company. While this is a noble profession, few people outside the industry recognize or appreciate the value this group provides to society. Many youngsters growing up dream of futures as a pilot, a doctor, a police officer, a firefighter, or an entrepreneur. Few likely aspire to be a refuse worker. To overcome this, senior management at this company went on a mission to elevate the pride of the workforce, calling the garbage workers "sentinels of the environment." At first, employees laughed. But it is true. "Imagine," said one manager to a large group of employees, "if you and the other thousands of garbage workers decided to stay home and not report to work for a month. What do you think the streets in your community would look like? Rats as big as cats would be running wild; restaurants, stores, and businesses would be unable to open." Management followed up these discussions by painting the truck fleet and applying each driver's name on his truck's door. They standardized uniforms and personal protective gear with the company safety logo. Over time, the employees started to view themselves and their jobs differently. They had a new sense of pride and accomplishment. Their uniforms were cleaner, their heads held much higher, they took better care of equipment, and they treated the customers and each other with renewed respect. Productivity was up, safety improved, costs were down, people were happier, and the company's bottom line benefitted.

Countermeasures: Overcoming Dysfunction and Turning the Culture

We have spent a good deal of text thus far outlining how to recognize and define dysfunctional practices. In this section, we'll address some of the techniques that we have successfully used to overcome and neutralize negative routines in reluctant, high-consequence companies.

Alignment around organizational realities—As we inferred above, the first step in healing a dysfunctional company is to recognize and agree on the market realities. Everyone in the company should align and be committed to chores that address the real issues. Unfortunately, as we pointed out earlier in this book, too often senior management ignores or wishes away the truth. Like children who manufacture their own reality while playing house, some managers play the same game by making up their own reality. That leads to "the myth of good." In these situations, as we pointed out, there is a reluctance to have candid conversations as the first step in overcoming barriers and challenges.

By contrast, the best-in-class companies consistently question what they do and why. They have regular dialogue and challenge each other to ensure their activities continue to translate into value-add for their customers. They deliberately seek out roadblocks and contradictions. They have candid conversations around market realities. They manage conflict, and they clarify roles. They agree on next steps and ensure everyone in the organization has clarity around where they are going, why it is important, and how to get there.

Leadership engagement. You can't delegate it—It is senior management's responsibility—even their obligation, to create a culture of engagement in an organization. An aloof, self-absorbed CEO and senior leadership team guarantees marginal results. It is a fool's folly to expect that midlevel managers and supervisors will take care of it. As discussed in chapter 4, there is no finish line to good leadership. You are tested over and over on your ability to lead people and promote alignment throughout the organization. Good leaders embrace the privilege of leadership. Bad leaders don't—they just hope for the best. Sam Walton, founder of Walmart and Sam's Club, said that, "outstanding leaders go out of their way to boost the self-esteem of their personnel. If people believe in themselves, it's amazing what they can accomplish." We like what we call our four E's of leadership: *Engage*: The best thing you can do to engage employees and leverage their goodwill is to demonstrate that they are important to you. *Execute*: Great operational and financial performance demands disciplined execution that drives quality processes. *Example*: Managers have to lead. Your people look to you. You are on display. Leading people from the front = winning. *Evaluate*: Good employees and managers want to be measured. Bad ones don't. Accountability elevates performance/results.

Middle management buy-in—Obtaining and sustaining support, alignment, and buy-in from company middle management may be your biggest challenge. It can also be your biggest bang for the buck. As the interface between strategy and execution, these people keep your organizational mission on track. They are closest to employees, customers, regulators, suppliers, and the public at large. They are the funnel for information flow up and down the organization. Top companies develop specific strategies around the care and feeding of their middle management. The poorly performing companies are ambivalent toward middle management. Taken for granted, and sometimes ignored by senior people, being a frontline manager can be a thankless grind. In one highly charged industry we know, the primary recruiting ground for entry and mid-level managers was from the front-line workforce. Traditionally, career progression was that good front-line employees would be invited to join the management ranks after a time. Today, however, it doesn't work. Front line workers don't always want the promotion because they see how their bosses are treated: consistently long hours, high-stress encounters with senior management, frequent family relocations, political games, risk of layoff on a whim, no overtime pay, etc. With this scenario, few in the ranks aspire to management jobs in their companies. This requires companies to recruit from the outside. Costs go up, spool-up time is extended, and management turnover is high because senior management hasn't fixed the problems that deteriorate manager quality of life.

Communications, messaging, and branding—Earlier we talked about human needs and the fact that you can't turn your needs off when you come to work. One of these needs is the desire to be part of something special. To be part of a group with special rights of passage where people can't just come and go at will. You have to earn your right to be part of it all. One of the more successful programs we managed was one where we branded the safety program by associating it with military fighter aviation and the space program. We conducted an internal employee campaign, inviting suggestions for logo and theme. Once we selected the logo and theme, we developed branding materials—hats, shirts/blouses, flags, stickers, etc. Then we instituted processes and prerequisites to certification—rules, testing, observations. Annually, employees and managers had to be trained and tested and achieve an 85 percent minimum passing score. All this helped create a mystique and people had pride in belonging, since not everyone is part of the group. Safety performance improved and casualty costs dropped. But that wasn't all. Productivity improved, pride in workmanship improved, people took better care of equipment, customer service improved, and the company made more money.

Dealing with internal skepticism—Good managers recognize that there will be some degree of skepticism when management introduces something new or different. Good managers also understand the self-interest nature of human interaction. That is, people need to understand how management actions and intentions will affect them individually. It is the "game theory"[2] we discussed earlier in this chapter. It is understanding this game that enables good managers to overcome and neutralize skepticism by including in the rollout strategy direct attention to the *How will this benefit*

---

[2] Theme identified by Len Fisher, author of *Rock, Paper, Scissors: Game Theory in Everyday Life*

me? message. People inside the organization will more readily buy in and support a plan or an idea if they see a direct-line benefit to them. This is part of the art of leadership.

Sense of urgency—A leadership display of positive energy with a sense of urgency will invigorate a team to action. Our experience has been that people want to be part of something dynamic. Most people want to see that their efforts are producing a positive impact. Smart managers mix a sense of urgency and aura of confidence to excite their staffs to action. The importance of high energy and a sense of urgency cannot be overstated. In creating a sense of urgency, people must be convinced that the company is in trouble and that none of this is optional.

Measure—create a heaven and hell—Good people want to be measured. Bad people don't. Good people take pride in producing results and demonstrating them with metrics. Bad people hide behind organizational "trees" and don't like measures that eliminate the hiding place. Great organizations measure performance, process, and execution. They create a "heaven" for those who succeed and a "hell" for those who don't. The good managers don't focus on motivating bad workers; rather, they focus on not demotivating the good ones. We have found that visible and applicable metrics move companies forward, highlight the good performers, and force attention to the right places. Lack of accountability is a culture killer. You simply cannot have a high-performing company without accountability. Promoting accountability is critical to ensure that people at all levels take pride in what they do and that they own the results. We know from chapter 2 that good companies maintain high standards and hold people accountable. Bad companies do not.

Sustaining success and building the mystique—To sustain positive movement, there must be visible and public recognition of people who do it right. Good management teams celebrate successes that uphold the desired values and principles when they come. We call these successes "cultural skid marks." Good managers promote and display the people who produce results the right way. With celebration of each skid mark, the culture seeps a little deeper into organizational DNA. The message in the organization soon becomes abundantly clear: *If you want to prosper in this company, you will buy in and engage.*

The preceding principles are foundational to understand what must be addressed in the transformational process. We expand on these cornerstone themes in the following chapters as we lay out how to do the actual work of change.

Chapter 6

CEO Involvement

Leading the Change and Engaging the Company

Chuck: As we saw in chapter 2, ineffective leadership plays a major role in destroying companies. The converse is also true: it takes dynamic leaders to transform a company in trouble into one that is on its way to great success. And as we saw in chapter 5, dysfunction does not fix itself. Chapter 4 laid out in generic terms the following main functions that are required to transform a company:

- Mobilize leadership/identify the need

- Build the team/build the plan

- Set expectations/follow through

- Expand ownership/institutionalize the change

There are many complicated actions within this list. Some involve problem-solving and building and implementing plans to fix specific issues within the company. Other actions involve learning about the company and its problems, engaging the employee base in getting involved, and then ultimately expanding the ownership of the transformation to include these very people. The CEO has to take the lead on all of this and is responsible for all of the parts. He will have a lot of very capable help, but the "buck stops" at the CEO's door. Only CEO involvement will make all of this happen in a bad company in need of change. However, while this leadership is absolutely key to the process, it is not sufficient by itself to get the job done. The CEO will define and lead the transformation effort but is still only one person. Unless the CEO can inspire ownership and involvement in the hearts and minds of the various levels of management as well as the rank and file, the effort to change and improve the company will only be partially successful— or could even fail miserably despite the best efforts of the leadership. Many transformation efforts fail because the CEO and his team fail to appreciate how important that second set of actions really is. The necessity to engage the rest of the employee base in rebuilding the company often proves to be trickier and more problematic than all the nuts and bolts that comprise a transformation program. When these efforts fail, it is usually due to people and not to processes or systems. This chapter will cover in detail what the CEO must do to start a transformation effort and make it be effective. Although the CEO will touch practically everything involving the transformation, the primary focus for the CEO will be the *people side* of the business. Only the CEO can engage the employee base, convince the rank and file that this change is for them, and effectively build an "army of change" within the company. We touched on this in chapter 4, but the real detail is here.

Owning the Transformation

In this book, we have seen that dysfunctional companies are very comfortable operating in failure mode. Taking a company that is content moving in one direction and getting it to start moving in another direction is hard work. Few people in dysfunctional companies gravitate toward any of this. They have to be strongly convinced by a very persuasive leader that this is worth doing. A CEO cannot get any initial traction unless he can demonstrate that his commitment to the effort and ownership of the effort cannot be swayed, will not be distracted, and will not go away. This is no place for a few speeches and then a handoff to someone else. When people see that kind of behavior, they will hand their piece off as well and the transformation becomes an orphan. The battle that will take place between the CEO with his team and the forces within the company that do not want change and want this effort to fail will be over the CEO's ownership of the transformation. The resisting forces are focused on destroying that ownership.

As we previously pointed out in chapter 4, John Kotter, in his book *Leading Change*, talks about defining the crisis and creating a sense of urgency. The opposing forces will ask, "What crisis?" Dysfunctional companies are experts at denial and rationalizing failure. The CEO will be greeted with myriad arguments that the company is actually pretty good, and big changes are not needed and would fail anyway. The quickest way to derail a call to change is to eliminate the need for the change right at the outset. The CEO must take on this "pushback" immediately and decisively by articulating the failure with strong facts. Bad companies are comfortable talking in terms of vague generalizations, but they hate facts. The case for change must be based on facts that are so obvious that even the apologists will be uncomfortable. What is important here is that the people in the company see that the CEO has done his homework, understands the crisis, and has a personal sense of urgency. Furthermore, the CEO has taken authority over the company and the improvements that are going to take place. There is no room for denial or any form of negotiation. There is no going back, and failure is not an option. This change is going to happen and people can either help or leave.

This initial commitment by the CEO is not an end in itself. This is just the opening salvo. Most of the management and the employee base in a dysfunctional company have been through these types of change processes before. They have heard the promises and the strong statements and then seen it all fail. Some have helped bring that failure about. CEOs often make the critical mistake of believing that their audience is not sophisticated and is hearing all of this for the first time. For some people this may be true, but for the majority, this is just another trip around the track. The audience is sizing up the CEO to determine if this effort will fail quickly or require a long time to be brought down. By stepping up, taking charge, and owning the transformation, the CEO is making the statement that he is formidable and is out to win. That will grab everyone's attention, which is a good start and will maybe get some of them thinking that perhaps this time it will be different.

The members of the audience may all be in agreement that change, accountability, and striving for success are not things they normally embrace. However, their reasons are not all the same, and it is vitally important that the CEO understands the various groups he is dealing with, why they act like they do, and what they want. There are typically five groups and they will have different responses to the ownership commitment.

- <u>The Good People</u>—These are the people who are saying, "Finally, a CEO who wants to do something. Maybe it will actually happen this time." These people basically have good values and want to work in a good company. They do not like the way things are but have seen so many promises turn into failures that they have a hard time believing it will succeed. In their work history, the CEOs have not been effective and have not inspired trust or faith. The initial ownership commitment can get them leaning in the right direction. But these people know how strong the internal forces of resistance are in the company. They will have to be continually reassured that the CEO and his team can prevail. The Good People will come on board, but they need to see some progress first. They can comprise a significant percentage of the employee base, but their members are generally passive. If they were action-oriented, they would have left already. Action has never worked for them. They do not embrace the prevailing dysfunctional culture in the company but see no effective way to oppose it. Now could be different. This group can eventually provide a number of the functional managers that will be required to implement the various pieces of the transformation. Many members of this group have been around for some time and know the internal workings of the company as well as the grassroots politics. If they convert, they will be seen as convinced disciples by much of the rank and file. The CEO has to connect with this group. Change will not happen if this group continues to sit on the sidelines.

- The Acceptors—Unlike the Good People, these employees do not see the upcoming change as good or bad. Many have been around a long time and are survivors. They will follow the company line whatever it may be. To them, the company provides them with a job that they want to keep. They view the CEOs, who have come and gone with their grandiose plans, with a lot of cynicism and amusement. The battles have been fun to watch. These people are not potential core leaders, but they are not usually trouble either. They always do what is in their best interests. If the CEO engages the Good People, the Acceptors will follow.

- The Rebels—This group of people represents the blatant defenders of the status quo. They are opposed to the entire message and are making arguments to anyone who will listen. The members of this group are often highly placed in the company and see themselves as so critical and essential that the CEO must tolerate them and listen to them. The CEO will be treated to several visits from senior members of this group where they will try to show him how this whole effort is so misguided. Some of these people are not very diplomatic and can be downright insubordinate. When this happens, it is a key event— because everyone is watching. If the CEO backs down, then the war is essentially over. In one example, a new CEO brought on to rebuild a company had to deal with an incumbent senior manager who was basically the leader of the Rebels. This individual was rude, uncooperative, not forthcoming with information, and generally an obstructionist concerning anything connected to the new agenda. After many public confrontations, the CEO did the unthinkable and terminated the senior manager. Everyone was in shock. This was a CEO who was not to be trifled with. The positive effect of this action lasted for about a year. At that time, another important Rebel became an issue and the CEO lost his nerve. Virtually all of these people need to go. A small few may be convinced by the CEO, but most who stay create a lot of havoc. It is best practice to remove the Rebels. To keep them is to send the message that their views and behaviors are acceptable.

- The Terrorists—This is a very dangerous group. They are the enemy. Many senior managers are in this group. They are as adamant in their opposition as the Rebels, but they do not show it. The Terrorists smile and nod and tell the CEO they are on board. In reality, they are veterans at opposing change. They have been misleading and outmaneuvering CEOs for most their careers. They are experts at killing initiatives and sabotaging just about anything that is generated by the corporate office. In their minds, CEOs come and go, but *they* will always be here and *they* are the real power in the company. The ownership commitment from the CEO does not impress them. They believe they have heard it all before. Many people know who the Terrorists are because they are very verbal on their own turf and when the CEO is not watching. The Terrorists are like cult heroes to their people who like to watch them finesse the CEO. When it becomes obvious that a new CEO is committed and formidable, the smarter Rebels become Terrorists. Much of this terrorist group has been around a long time and has won many battles. They plan to win this one as well. The transformation effort can only succeed if the CEO and his team can neutralize this group and their efforts to obstruct. Too many CEOs make the mistake of trying to win these people over to the effort. That is a recipe for failure. The best course of action is to determine who is a Terrorist and who is actually in the next group (a neutral observer) and remove the Terrorists from the company as quickly as possible.

- The Fence-Sitters—This group is similar to the Acceptors, but they eventually take a position. They initially are neutral and watch the battle. They always side with the winners. The Fence-Sitters are not bad people. Some of them really care about the company and would like to be part of something successful. They are just risk-averse and have seen so much transformation failure that they do not have much trust in the company leadership to be able to pull it off. They are not people with a lot of faith. They want facts and results. This group is large, and how they respond is a great indicator of how the transformation is going. Most are born followers, but there can be some reliable managers in this group. The problem is that they do not show up until later, if at all. This group will have interest in the CEO's commitment to ownership of the transformation, but this will not move them out of neutral. They know that the Terrorists usually win these duels, but they will be watching to see if this CEO is different and has the tenacity, intelligence, and leadership skills to be successful in the face of very capable resistance. The CEO commitment is at least a good start and has their attention. Who wins this group will ultimately win the battle, so the CEO must understand these people and connect with them.

It should be obvious by now that a CEO brought in to transform a bad company cannot be your average company executive. Most CEOs will lose heart and wilt in the face of this much organized and seasoned opposition. The Terrorists and the Rebels know how to bring these weak and uncommitted leaders down. A new CEO with a turnaround plan pretty much comes into the company alone. Some of the employee base are strongly opposed to his plans, many are neutral, and a very few are supportive. And even this last group cannot be completely trusted or relied upon to be very helpful. The CEO has to either bring his team with him or cultivate members from a very narrow group of supporters. To be successful, the CEO has to be tougher, smarter, more tenacious, and more committed than the people who are there to oppose his efforts at change. That initial commitment to ownership marks the start of the game. The CEO will not win with it—but the other side knows he is serious. Going forward, the CEO has to stay true to that initial commitment. Without that, even a good team with good plans will become isolated and fail.

At this point, the reader is probably wondering why the Rebels and Terrorists are not just eliminated right at the beginning. That would be great if it were feasible. Removing people, just like bringing new people in, is not a mass exercise. Each step must be carefully planned, done for the right reasons, and done correctly. And most importantly, done one person at a time. The work effort and time commitment are significant, but there are no shortcuts here. Additionally, most Terrorists and Rebels do not step up and present their credentials. They have to be identified and then scrutinized, and that also takes time. It is safe to say that a transforming CEO will be dealing with Rebels and Terrorists almost every step of the way and should never underestimate them. The important thing is that the group will be getting smaller and the diligent CEO will be keeping them on the defensive.

Building the Plan

After the initial commitment, building the transformation plan is the next major step. As we saw in chapter 4, building the team is the initial step in building the plan, so it goes without saying that the CEO is just as focused on who builds the plan and how they build the plan as what is in the plan. The actual development of the plan is covered in chapter 7. This section deals with the role of the CEO. The intent here is to accomplish a number of very basic things that are strongly needed to keep the program moving forward and build support within the manager and employee ranks. The following steps are key, and only the CEO can initiate these things:

1. Continue to demonstrate strong CEO commitment

2. Start developing some rapport and initial feelings of ownership within the manager ranks

3. Build some traction with employees who are neutral or mildly supportive

The initial message must be that the CEO will get this plan completed and implemented, and that the content of the plan will have the necessary quality to move the company. The CEO should appoint a team of strong leaders to take the program forward, but it must be clear to everyone that the CEO will stay involved to make sure proceedings meet his standards. There is no room here for a disconnect between the CEO's initial message and the final product. One must flow into the other, and those watching must be impressed with what they see. The need for any damage control at this stage in the plan will push people away and embolden the opposition.

A key message early in the process is that the CEO has no intention of importing some "off the shelf" program and then pushing it down on the company. The plan will be custom-built by a coalition composed of the CEO's team and employees from within the company. It must be clear that the CEO wants input and involvement from deep in the organization. The plan must address the needs of the company. The CEO has to walk a fine line between authority and collaboration to ensure the plan has quality and meets his standards but is not seen merely as an edict. The intent here is to create some sense of ownership within the development team and focus on building something that they can see as a positive accomplishment. It is vitally important here that the CEO establish at least an initial beachhead of trust and rapport with the managers and employees selected to participate in the plan development. Others will be watching and their future attitudes and actions will be affected by what they see and hear from these people.

Another critical component that is often overlooked is keeping employees in the loop as the plan goes through development. For some reason, CEOs often make this a very secretive process and only people with special clearance get to have a look. When there is communication, it is highly sanitized and more of a commercial than a real update. This approach sends the message that employees are not worthy to know and makes everyone suspicious about what is really happening. People are driven away by this tactic—which is just the opposite of what is needed. This is a great opportunity to engage employees by bringing them close to the process and being candid with them. They are not used to this type of fair-handed treatment from management, and it will get their attention. The CEO should take extra pains to keep everyone up to date on progress on a regular basis. The communication should be often, timely, and truthful. Accomplishments, as well as problems and surprises, should be shared with the employees. If these messages can get people invested in the success of the program to the point that they offer unsolicited ideas and suggestions, so much the better. The main thing here is to get their attention and make them feel involved.

The broad coalition that actually crafts the plan is a key component to the overall process. Chapter 4 describes in a generic sense what types of people the team should include and the reasons for that. The CEO should select a team that is composed of a mix of new people brought in by the CEO as well as incumbent managers and employees. The team should draw from all the various parts of the company as well as from top to bottom in an effort to cover most of the possible viewpoints. The one common denominator is that all team members should be capable of working together to constructively build a quality transformation plan that meets the standards of the CEO. They must be able to do the job.

There are numerous reasons for the CEO to interact with the broad coalition while they do their work. The CEO is still new to the company and does not really know anyone very well. Furthermore, the CEO is pretty much an unknown quantity to the company employees, other than the few public statements that he has made concerning the transformation. Working with the various subgroups within the broad coalition is a great opportunity to see people function firsthand and learn more about them individually and as a group. The following items are some of the things that the CEO should be focusing on as he moves from subgroup to subgroup and interacts with the team members:

1. Learn more about the attitudes and abilities of incumbent senior leaders

2. Find untapped talent within the company

3. See how the basic culture within the company affects action

4. Get more of a feel for the general response to the transformation

5. Identify the major issues in people's minds

We know from earlier in this chapter that incumbent senior leaders can have a range of perspectives and that many are well versed at concealing attitudes and manipulating CEOs. An accurate assessment of these people can require a lot of observation time and considerable interaction and investigation. The CEO has to determine early on who can be trusted with leadership roles in the transformation, who needs to be watched, and who needs to go. The plan development period provides a good opportunity to get this evaluation process of senior people started.

Dysfunctional companies are typically not good at identifying, using, or rewarding talent. Consequently, spotting talent in these bad companies is not very easy. The leaders and managers are not necessarily the most capable people in the mix. The broad coalition gives the CEO and his team a chance to identify people within the organization who can contribute positively to the plan and possibly be leaders in taking the plan forward. The CEO should be looking to stack the broad coalition with as many of these people as he can find and should add more later as they are identified. An additional benefit is that these people will see the CEO and the transformation as personal opportunities to advance in their careers, and this will get them committed to helping improve the company.

Even though this is a new transformation effort involving a mix of people who have never worked together before, the existing culture within the company will show itself as people start to interact and deal with issues. This is a good opportunity for the CEO to see firsthand what the baseline company really looks like in the trenches. The more dysfunctional routines and attitudes will help target priority areas to be addressed in the transformation plan and will create teachable moments for the CEO and his team. The coalition members can be shown with real-time examples where they are and where they need to go. There may even be some good practices that can be brought forward and retained in the new program.

At this point, the CEO has developed an initial feeling about the company, including apparent gaps in its culture, prevailing bad habits, and basic management and leadership issues. This has all been based on what he has read or been told, and perhaps from interviews of selected senior people within the organization. The picture is far from clear or complete, but it has served as the basis for the CEO's assessment of the situation and what the transformation, at least in general terms, should address. Now is the time to get more specific input from a broader base within the company regarding the condition of the company, responses to the transformation effort, and what people see as the critical issues and obstacles. The coalition group and the work they will be doing provide a perfect opportunity to get this information. In addition to making these questions part of the actual work agenda and listening to comments during the work sessions, the CEO and his team should be getting more feedback in sidebar conversations and casual one-on-one opportunities before and after the meetings and during breaks. The intent here is to get people comfortable talking to senior leaders and expressing their views. Candid feedback is an absolute necessity during this entire transformation effort if the CEO wants to know if the program is correctly focused, on track, and striking a chord with people.

Plan development must move at the correct pace. The Terrorists and the Rebels will try to delay it, slow it down, and sabotage it. They know that time is on their side. Unless the CEO continues to push and can show steady progress, the natural resisting forces within the company will grind the program to a halt. In bad companies, improvement efforts that have been stopped virtually never start up again. The people, from top to bottom, lose heart quickly and then move away. Although the CEO must keep pushing so that he has a work product to show everyone while people are still interested, it is also important to take enough time to build a quality plan that will be effective during implementation. A substandard program that is ahead of schedule will not work here. The CEO has only minimal credibility at this point in the process and cannot afford to put out a bad product. The program should not be released in final form until the CEO is confident that his team understands the key issues within the company, has the plan correctly focused on these priority issues, and knows that the various transformation modules are feasible and can be mastered by managers and their employees.

If the CEO handles this development phase of the program correctly, the transformation team should go into the next phase with higher credibility, more people believing in the program, and greater confidence that they can engage this company and be successful. The CEO will not only have a plan that meets his standards but will also feel that he knows a lot more about this company and its managers and employees.

Communicating the Plan

We touched on the communication piece in chapter 4, but much more needs to be said about this topic. First of all, we are dealing with an employee base that did not choose this direction for the company. As a group, they are wary of change and a lot of change is now going on around them. At best, they are going to be hopeful, but more likely the attitude early in the process will range from skeptical to horrified and angry. In their careers, they have heard many promises from senior management, but few have been kept. System failure and surrender have generally been the normal outcomes for any movement to improve the company. CEOs often make the mistake of dismissing the importance of employee attitudes in the whole transformation process. They think people will embrace their ideas and edicts simply because they are the CEO or, at any rate, people do what they are told. This is exactly what the Terrorists want from the CEO. When people are pushed into doing something by a CEO who has no great desire to connect with them, the Terrorists can easily show them effective ways to resist and ultimately kill the plan. The CEO must understand up front that either he engages the employee base with a strong communications effort or the Terrorists will take over that job.

Initially the transformation plan belongs to the CEO. The initial commitment states that fact, and no one else really wants it anyway. For the plan to be successful long-term and become an integral part of the fabric of the company, the CEO must focus on transferring some of that ownership to the employees. However, the CEO cannot make that happen. The employees have to decide that they want it to happen. This is where the communications effort comes into play. The CEO and his team must continuously provide good reasons to the employees for moving in that direction. The balance here is critical. The communications effort must be convincing, but it cannot look like a sales campaign by an ad agency. It must be fact-based, truthful, unbiased, and interesting. Employees are primarily concerned with their interests and, perhaps secondarily, the interests of the company. The messages must show how the transformation is the best thing for them, their career, and the company. The arguments must be easy to understand and present points that even the most skeptical listeners cannot dismiss. It is vital that these messages connect with the audience.

Effective communications efforts, including TV commercials, have a spokesperson that becomes synonymous with the message. It can be a duck, a talking reptile, a guy with a deep voice, or a clown named Jack, but people take notice when these spokespersons appear, and they pretty much know the message before they hear it. And for some strange reason, they believe that message. In this case, the spokesperson is the CEO. The plan is working if every time employees see the CEO in person, in a video, on a poster, or in a brochure or other written document, they immediately think about the transformation and all the key messages connected to it. To make this effective, the CEO has to step up to the role of spokesperson. What that means is being consistent, staying on message, having enthusiasm for the message, and making a genuine effort to connect with the audience. The CEO has to "pitch a no-hitter" here, because the Terrorists will be looking for signs of weakness. If they can damage the image of the spokesperson, then they can damage the program. Conversely, a spokesperson who comes out of the process unscathed by the opposition becomes a virtual icon for the transformation. That credibility is priceless and produces great results. Gordon Bethune, who led the turnaround at Continental Airlines in the 1990s, was a leader who was very good at interacting with the rank and file. He knew his people and he knew their jobs. When he instituted a plan to dramatically improve on-time performance by getting the various functions to work together on this single goal, he became synonymous with that program and the positive impact it had on the operating culture at Continental.

Now that we know who sends the message, what is in the message? First off, the CEO needs to reiterate his commitment and enthusiasm for the transformation. People have to hear things six to ten times (we know this from commercials on TV) before they register, and this is a message that needs to register if everything else is going to happen. Employees have to believe that the CEO will be with them every step of the way and will make sure it all comes to fruition. The CEO next should get the audience comfortable with the idea of change. They have a natural fear of change and need to hear believable arguments on why change can be their friend and make their lives and careers much better. This message must not only connect with managers but with employees at all levels. They all must see that the plan has something in there for them and that there is a good reason to get on board. Managers and employees have been buying the argument for years from the Terrorists and the Rebels that the environment of chaos and "no rules" in which they have been living is best for them. This argument cannot stand up to any level of rational scrutiny and only works on people who are not thinking, not paying attention, or have become passive. The CEO has to bring everyone back to reality by confronting this weak argument at every opportunity. It must be clearly shown with examples that good people work best in an organized environment where there are strategies and processes as well as goals and accountability for those goals. Only the dysfunctional people want to maintain the chaos, because it gives them control to the detriment of the company and everyone else. The CEO must offer an alternative to the people based on facts and not fear and lies. The CEO should conclude the message by stating that *he wants to hear from the employee base*. Virtually all CEOs say this, and the people know they do not really mean it. This has to be different. If the CEO is going to engage these people and get them to trust and believe in the company leadership, then he has to give them a chance to speak. The CEO and his team need to organize feedback forums across the employee base, where groups of reasonable size can ask questions and voice concerns on a regular basis. This is not the favorite activity of most CEOs, but it must be done if all of this is going to work. The CEO should view it as another key task in the overall effort to engage the employee ranks and make the transformation successful. The message is about commitment and trust, and concluding with a request for dialogue really communicates.

The communications plan must be cognizant of all the groups within the company. While the target groups are the Good People, the Acceptors, and the Fence-Sitters, the Rebels and Terrorists are carefully watching and looking for chances to offer rebuttal arguments to the masses. The goal is to consistently present high-quality messages based on facts, supported by examples that will steadily convince the target groups that the transformation is good for them and will be successful. At the same time, the message to the Rebels and Terrorists should be that their world is changing and their way of doing things is going away. They must either change or leave.

Setting Expectations

Once the transformation plan is complete and meets the CEO's standards, it is ready for implementation. Details of the actual implementation are covered in chapter 9. What are addressed here are the messages that the CEO needs to deliver just prior to and during implementation. Up to this point, the CEO message has focused on generalities, such as: why a transformation is needed, the advantages of a transformation, and what a successful transformation looks like. Now is the time to start dealing in specifics. The employee base needs a very clear picture of what will happen, what will be their role, and what will be the outcomes. They are not used to this much granularity from the CEO position, so this is a key event in the overall process. It will get their attention and start them thinking about success as a real attainable thing and not some concept that is always out of reach. It is vitally important that the CEO be on target with the initial message regarding expectations and then to stay on track with additional messages as the company goes through the entire transformation effort.

The CEO should start by reaffirming his commitment to the program. This may seem redundant, but the audience needs to keep hearing it. It is the one point that will connect all CEO messages from beginning to end and is the primary basis for much of the support the CEO has in the ranks. Next, the CEO must define his role in the implementation phase. The managers and employees who have an interest in the transformation want to know that the CEO will have a hand in taking the program forward, and they want to know what that role will be. The CEO can get as specific about his role in the implementation plan as he feels is necessary, but the message must include, as a minimum, the following three points:

1. The CEO is proud of the transformation plan and wants to be identified with its implementation.

2. The CEO wants a very specific and productive role in the rollout of the plan.

3. It is important to the CEO that the audience is well aware of his involvement and knows what to expect from him.

These three statements all have a common theme, which is that the CEO's involvement in the process is much more than ceremonial; there is real work involved. Do not forget that the continuing battle between the transformation team and the Rebels and Terrorists is over the CEO's ownership of the transformation. The employee base needs to keep hearing that the CEO still holds that ground. Furthermore, the CEO is the spokesperson for the plan and is the one person the employee base sees as critical to its success. If the CEO is ready to roll up his sleeves, then it is time for everyone to do the same. This is a call to action.

The next step is to present the road map of the implementation for the whole company. Clarity and detail are important. The CEO should go over the timeline, important tasks, benchmark events, and who will have leadership roles. The message should contain enough detail to demonstrate that events are well planned and the program is on the road to success. The CEO should also present the program goals, which are basically the CEO's goals. Since the incumbent company culture is not big on planning and outcomes, this will be a watershed day for many people in the company. The motivation here is to get them thinking differently, not only about the transformation but about their view of the company and its capabilities and behaviors going forward.

The message now must move toward building ownership attitudes for everyone by getting specific about their roles and their goals. This is an opportunity for the CEO to talk about good company cultures where everyone gets involved and is accountable for what they accomplish. This is not only a teachable moment for most people as they find out what their role is in the transformation; it is a fork in the road for the remaining Rebels and Terrorists, who are also challenged with specific assignments and responsibilities. These opposition groups are not used to being pushed back on their heels and having to react publicly to a call to action by the CEO. The CEO must make them decide which way they will go. There is no room here for compromise or special treatment.

They must either change and commit or find a new place to create havoc. This is a new day for them as well.

A big part of the transformation plan involves accountability. The company has never accomplished much because there has never been any ownership or accountability in anything that they have ever tried to do. That has to change if this program is going to be successful, so this is a critical step for the company. The CEO has talked a lot about ownership, along with commitment and trust, and the rank and file is probably starting to get the message. However, commitment without action is not enough. Now is the time to start talking about action and the accountability that must accompany it. The managers and employees have been shown the road map for the plan and they all know their roles. The company must now make the next step. Ownership and commitment put into action produces successful outcomes, provided there is accountability for that action. The review and follow-up process is really a maintenance function, but it must be presented here to help people see the crucial connection between action and accountability. The CEO must stress that staying on top of performance and making necessary corrections is how good companies do great things, and the review and follow-up process is the key that opens that door. This will be a leap of faith for the employee base, but a leap that they must make. Their history has been to avoid accountability and see it as a bad thing. The CEO will have to use all his credibility and goodwill to make this happen. The review and follow-up process should be started early in the implementation phase as a teaching tool with the CEO personally involved. The managers and employees will have to learn by example that accountability will make them better and that working without it is a recipe for disaster. Maintaining standards is critical here, so there will be what can be construed as negative feedback to managers and employees. That has to be handled masterfully so that it becomes a teaching opportunity and not a morale buster. The biggest mistake the CEO can make here is improperly managing the early reviews or delegating them to others, who then fumble the process. There is no room here for mistakes if the employees are going to be convinced that reviews are good for them.

The CEO must complete the expectations message by defining success. This involves describing what will be accomplished with the transformation plan and what the company will look like long-term. A critical mistake here is to oversell the vision and promise something that is a reach, or not even feasible or believable. The employee base must be able to see the long-term vision as something that is positive but also attainable. The communication should have enough detail that the audience can connect the various modules and tasks to the resulting accomplishments. On a personal level, each employee must be able to see what will be the outcomes from the new things he is being asked to do. Up to now, the CEO has been the spokesperson and an icon for the transformation. The CEO will always keep the spokesperson role, but with time, the vision of success must become the icon. It is the whole reason why all this work is being done. The CEO should continuously provide progress reports that not only celebrate successes but talk about setbacks and the corrections that are being instituted to address them. The intent here is to keep the transformation present in the minds of everyone and get them emotionally involved in the whole process. They need to know that the CEO wants them informed, and in kind they need to want to be informed because it is important to them. This is also covered in the next section, *Program Maintenance.*

Program Maintenance

The most important thing the CEO can do during implementation is to be visible. The rank and file associates the transformation with the CEO and expect to see him actively involved on a regular basis. When the CEO is present everyone is more upbeat, more confident, and more focused on making the program move forward. Although the CEO cannot micromanage the program and has to let the program leaders operate and build their own credibility, more will get done with less confusion when the CEO is in the room and paying attention. A smart CEO can be in the room and definitely in charge but not dominating everything that is happening. Staying in control but letting others do their jobs is the key.

It is vitally important that the program stay on track, and the CEO is the best person to make that happen on a regular basis, especially early in the implementation phase.

The CEO needs to cover a lot of ground during implementation. Although other members of the CEO's team are directly responsible for the various tasks and modules, it should appear to the employee base that the CEO is everywhere, touching everything and knowing everything that is worth knowing. The CEO should be interacting with many people at all levels and challenging them to step up and do their part. It should be clear that the transformation is the number one priority to the CEO and he is not getting distracted. The CEO is the spokesperson for the program, and that role is needed now more than ever. Obviously, the CEO cannot be at every meeting, but he should attend as many as possible. Priority should be placed on meetings involving critical issues or difficult participants, and those being led by less effective team leaders. The CEO must cultivate team leaders who have the potential to build their own credibility and rapport with the employee base. This will help to broaden the CEO's influence during implementation. The Terrorists will be trying to sabotage proceedings and may actually challenge some of the task and module leaders. They will be less confrontational with the CEO around and it will be important for the rank and file to see them neutralized. The CEO must make it clear that he has control of this company and troublemakers will not be humored or tolerated. Anyone who dares to challenge the CEO should be dealt with quickly and publicly. One clarification must be made at this point. Challenging the CEO does not mean having constructive questions or making suggestions that are heartfelt and well-thought-out. A CEO who cannot deal with this type of input has the CEO disease. Unacceptable challenging involves being obstructive and trying to undermine the overall effort rather than trying to improve it. This type of input is not needed and must be stopped right at the beginning.

Not everything will go well during implementation and some basic assumptions may even prove to be incorrect. In these times, dysfunctional companies tend to become paralyzed and lose heart. They are not good at fixing their problems. The CEO can have a powerful influence during these difficult times. The CEO is the one person who can take control of the situation, make sure issues are being dealt with properly, and assure people that these obstacles are not insurmountable. The connection between commitment, action, and accountability is now more important than ever, and the CEO is the one to get that point across. Although issues can destroy a program if not understood and dealt with appropriately, when issues are handled successfully, the program team and the company as a whole become more confident and stronger. The CEO should also involve other members of his team in fixing things so that the overall capability of the team can grow, but the key here is effectiveness, so the CEO should transfer control very carefully.

The CEO should personally issue progress reports on a regular basis and conduct program reviews at important times such as the attainment of key benchmarks. The intent is for the CEO to stay visible, keep people informed on progress, and repeat the key messages on commitment, action, and accountability. On all of these occasions, the CEO should be open and candid with the audience and continue to tell them that they are partners in this enterprise. These are opportunities to celebrate successes, demonstrate the growing ability of the company to deal effectively with its issues, and to get people believing in themselves and the company as a whole. The enemy here is complacency. It is common for companies to get used to the success and dial back the effort. The CEO must end every message with the reminder that all this success came through commitment and hard work. It can easily disappear if the commitment and hard work go away. The old company was not up to the challenge—but the new company is.

As this chapter has shown, the CEO's role in the transformation is massive. In the face of significant opposition and an unconfident employee base, the CEO is building a brand, and that brand is the new company. To get the job done, the CEO has to communicate a vision of the new company, engage the rank and file in that vision, convince them that they are up to the task, and then convert them to taking ownership in the effort to rebuild the company. As was stated in the first chapter, dysfunction grows quickly and naturally, while organization and success require a lot of commitment and continuous attention and hard work. The CEO and his team must be ever mindful that they cannot take their foot off the gas and they must keep communicating that key message to the rank and file. Positive transformation is a never-ending process that has a beginning but does not have an ending.

Chapter 7

Building the Team

"Individual commitment to a group effort—that is what makes a team work, a company work, a society work, a civilization work."

—Vince Lombardi

Jim: Hire for attitude. Train the skills. This is the standard litany from many current management books. Good companies need people who have the appetite to be part of a well-functioning team. Managers in these companies look for every edge they can find to improve a competitive advantage. Smart managers understand that good relationships within the team get better results. They know the positive impact of a solid team ethos, where there is alignment around values and clarity around roles, expectations, and deliverables. The effort is reinforced by the right metrics, back-office support systems, and conscious management efforts to maintain alignment.

Getting the Right People

In this chapter, we expand on some key points on the importance of building a motivated and aligned team. In chapter 2 we emphasized that the first step in the people process is building a solid team. The best companies work tirelessly to execute the fundamentals with precision. In a December 4, 2018, article, "Why Southwest Says Soft Skills Reign Supreme," posted by Amy Elisa Jackson on Glassdoor.com, Greg Muccio, Director of People at Southwest Airlines, addressed the point. "I challenge our recruiters to not be afraid to present to a hiring department a candidate who may not have all the hard skills they want, but who truly displays all of our core values. We can train for skill, but attitude is a bit harder to teach." He goes on to say, "While other companies may place technical skills over traits such as conflict resolution and dependability, for Southwest, those soft skills are essential. ....I would actually call them essential skills ....some of the most crucial of these skills include: communication, teamwork, leadership, relationship-building, balance, reliability, and dependability."

As we said, dysfunctional companies don't do this. They either ignore or botch this critical step. Bad leaders don't value talented people and put little effort into who they put on the team—it is too much work to develop necessary quality processes to ensure the team has the right players in the right seats. Bad leaders freelance, shoot from the hip, and usually select whoever happens to be convenient, with little or no thought to define what they need to nurture an effective team. They are reactive and want to fill the vacancy as quickly as possible. There is no thought given to assessing skills, force ranking, or evaluating people and replacing those not capable of delivering necessary results. It is easier in bad companies to keep mediocre people who don't challenge the status quo. This attribute is especially prevalent in industries that believe that people from outside their industry bring little value. We have been part of companies that believe it is better to go with an unexceptional performer with a history in the industry than it is to bring in a stellar performer from outside the business. This is a flat spin and guarantees continued second-rate organizational performance.

On the other hand, good leaders and good organizations recognize that they must develop the right architecture, including defining the qualities, skills, temperament, and cultural fit they need to take performance to the next level. They look not just at what they need today but at what they will need tomorrow to continue to excel in the marketplace. Good leaders know that anchoring the right processes and positive atmosphere in the organizational bone marrow is what advances and sustains solid team character for the long haul and determines who wins and who loses on the business battlefield. It is about leading people and accepting the reality of human dynamics. Failure to master that reality can result in dysfunction and marginal performance over the long haul.

The Guiding Coalition

Sometimes major transformations are associated with one individual, such as Gordon Bethune, Lee Iacocca, Sam Walton, and Jack Welch. And while one strong individual can initiate movement, sustained major change is so difficult in most organizations that it takes a powerful group to sustain and institutionalize the change. It is what Jon Kotter calls a *Guiding Coalition*. We can attest that any change effort needs a guiding coalition to move the transformation forward. In our view, this is where the magic can happen.

In chapter 4, we detailed some of the benefits of a guiding coalition. Getting the right people on that coalition and empowering them to engage the workforce and neutralize the Terrorists and the Rebels is essential to drive good change. Some characteristics of an effective guiding coalition include:

- Members must have energy and passion and be fully aligned with the CEO's vision.

- Members must carry and reinforce the message to the front line.

- Members must actively engage the rank and file throughout the transformation.

- The coalition must operate outside the normal hierarchy and report to the CEO.

- The group must have credibility in the eyes of everyone in the organization.

We both have directed guiding coalitions while engaged in leading culture change in major organizations. In general, here are a few lessons we learned about ensuring coalition effectiveness:

1. The coalition must be led by a credible, high-energy, positive leader.

2. Identify and recruit *thought leaders* in the company for the coalition. They must be people of influence and power as opposed to people with mere authority by virtue of rank. We always included people in the trenches—frontline workers, midlevel managers, supervisors, employee representatives, and labor union leaders. The coalition must have a mix of leaders who can orchestrate the change, drive the creative process, and deal with the detail and potential chaos that may be generated by the transformation.

3. Spend time to nurture the team and ensure that they are aligned with the CEO's vision. Show them why this is important to them and why it helps them in the long run.

4. Try to keep the coalition to a manageable size. Once the word is out, there will be many who want to be part of it.

5. In big organizations, you will likely need to assign subcommittees to address specific tasks and then report back to the central coalition.

6. The coalition should operate outside the normal hierarchy and report to the CEO. To help avoid paranoia by some in senior management that the group is somehow scheming against them, remove the mystery of coalition actions by keeping the upper echelon informed regularly. This is part

of the art of change management, and the coalition leader needs to be on top of this one.

7. Managing the coalition and the process is critical. Ensure regular, consistent, and direct communications with coalition members to keep engaged. This is especially important if coalition members are located at various locations. They need to know that the work is ongoing and impactful. This isn't a time to assume the team knows what is happening.

Companies that fail usually underestimate the difficulties of producing change—that is why the guiding coalition is so important.

Power to the People

We know that one of the essential elements in building an aligned and effective team is to make certain that the frontline workforce goes along with you on the journey. Ultimately, frontline workers are the ones who deliver the service and bring in revenue. Senior management depends upon the workforce to fulfill their commitments to customers. We don't know of many senior executives who directly pick up the trash cans, or switch a box car, or write an invoice, or load a bag on an airliner, or wait a table, or patch a hole in the street, or any number of millions of other tasks that must be done every day to serve customers and fulfill an organization's mission. Quite literally, a CEO can be virtually powerless, even while holding great authority. Sure, a CEO can order that things be done, but if the people on the front end decide not to do a job, or to do a job poorly, what does a CEO do? Fire everybody? Is that a reasonable option? Does it ensure that things are now right with the customer who was disappointed? Does it address organizational gaps that may have led to the failure? The reality of leadership is that CEOs must influence the incumbent workforce to execute the plan as a team if they are to win in the marketplace. Successful leaders know that, in most circumstances,[3] persuasion,

---

[3] Of course, there may be times when edicts are required to address immediate high-consequence business needs when there is no time to confer.

inclusion, and engagement work much better in building a productive team than issuing edicts to the workforce from the corner office. Horst Schulze, former president of the Ritz-Carlton Hotel Company, was quoted as saying, "We are superior to the competition because we hire employees who work in an environment of belonging and purpose. We foster a climate where the employee can deliver what the customer wants. You cannot deliver what the customer wants by controlling the employee."

Nothing will deter and crater a change effort more quickly than charging ahead and not seeking frontline team input. The guiding coalition can help in very large companies, but the CEO needs to have a clear communications strategy that includes direct contact with employees to reaffirm values and vision and keep people informed on the journey. If you have a vexing problem, go to the workforce and solicit direct input. Let them help you solve the problem. Be sure to communicate back to the team the implementation strategy, the details, the actions, and the whys.

So what does it take to build a successful and connected team? Many of the chapters in this book examine fundamental traits necessary to build and support strong team character and positive corporate culture. Building upon many of these qualities, in this chapter, we offer expanded thoughts on several themes that can affect group performance. Senior managers who practice these principles have greater success in keeping their teams unified around the mission than those who do not understand the significance of these concepts.

Leaders vs. Managers—Which One Is Best?

> "Management is doing things right. Leadership is doing the right things."
>
> —Peter Drucker

In chapter 1 we introduced the concept of Leaders vs. Managers. We said that in bad companies, there is very little thought given to how the roles relate to each other. To expand the concept, we have found that in order to build and sustain durable team character in any organization, there must be clear executive understanding of how leadership qualities link with management qualities. One of the questions that warrants discussion at this point: Leadership or management—which quality is best? We know from experience that this topic can generate spirited conversation in corporate halls and academic circles as to which attribute is most desired in organizations. We have seen instances where some consider management qualities as somehow less important than leadership, even to the point of considering managing a lesser discipline.

Undeniably, leadership is different from management. We are partial to the way John P. Kotter[4] explained the relationship between the two qualities in a 2001 *Harvard Business Review* article titled, "What Leaders Really Do."

What managers do

> "Management is about coping with complexity," Kotter wrote. "Without good management, complex enterprises tend to become chaotic in ways that threaten their very existence. Good management brings a degree of order and consistency to key dimensions like the quality and profitability of products."

What leaders do

> Leadership, by contrast, is about coping with change. "Part of the reason it has become so important in recent years is that the business world has become more competitive and more volatile [and] the net result is that doing what was done yesterday, or doing it 5% better, is no longer a formula for success. Major changes are more and more necessary to survive and compete effectively

---

[4] Now retired, **John P. Kotter** was a professor of organizational behavior at Harvard Business School.

in this new environment. More change always demands more leadership," explains Kotter.

Why You Need Both

Bottom line, Kotter concludes, is that "leadership is not necessarily better than management or a replacement for it. Rather, leadership and management are two distinctive and complementary systems of action. Each has its own function and characteristic activities. Both are necessary for success in an increasingly complex and volatile business environment. While improving leadership abilities, companies should remember that strong leadership with weak management is no better, and is sometimes actually worse, than the reverse. The real challenge is to combine strong leadership and strong management and use each to balance the other."

Our belief is that good organizations recognize the value of a cadre that includes both leaders and managers. Leaders provide the vision for the team, and managers provide the structure upon which the vision can be achieved. In dysfunctional companies, executives don't recognize, nor do they value, the criticality of having people who are both good leaders and good managers.

Positive or Negative—It Is Up to You

"Half the harm that is done in this world is due to people who want to feel important ....they do not mean to do harm ....they are absorbed in the endless struggle to think well of themselves."

—T. S. Eliot

As we alluded earlier in this chapter, building and sustaining a successful team culture is hard work. It requires managerial commitment and focus. People in the organization have the power, and that power can be exerted in a positive or a negative manner. Which direction the culture goes depends upon the atmosphere the management allows to develop. Good managers and leaders recognize and address negative issues consistently and quickly. Mediocre or poor ones ignore or are unaware of the gradual negative cultural creep and, in time, find themselves leading discordant teams. Following, we have highlighted some of the things that we have seen that can affect team effectiveness if left unattended by senior management.

Toxic People

In chapter 6 we listed five types of people most commonly encountered in organizations: 1) the Good People, 2) the Acceptors, 3) the Rebels, 4) the Terrorists, 5) the Fence-Sitters. In this section, we want to expand our discussion on the Rebels and the Terrorists. These people are toxic, scoring high on the Machiavellian scale characterized by a deceitful interpersonal style and a focus on self-interest and personal gain at any expense. Left unchecked, they can make a healthy company sick by robbing emotional and intellectual vitality from the good people. In a November 2016 *Harvard Business Review* article, "Isolate Toxic Employees to Reduce Their Negative Effects," Christine Porath noted: "The evidence is clear: having a toxic or de-energizing employee on your team or elsewhere in your organization's ranks is costly. Not only are the people around the person negatively affected ....the people close to a toxic employee are more likely to become toxic themselves."

We have witnessed destructive traits injected into a company's soul by toxic people because they were tolerated by weak-willed managers and leaders. Here are a few examples of the more common destructive traits that we have seen exhibited by Terrorists and Rebels that destroy team vitality:

1. <u>When the meeting ends, the treachery begins</u>—A Malayan proverb says: "Don't think there are no crocodiles because the water is calm." That applies here. Let's say you have a cordial, cooperative meeting where issues are raised and discussed and decisions are made. Everyone around the table nods and ostensibly agrees. You have a plan, right?

Well, not if there is a Rebel in the room. Despite seemingly in support during the meeting, once the meeting ends, the Rebel does an end-run around established forums by "storming outside" with covert resistance to discredit and undermine the team.

2. <u>Blue bloods</u>—These are people who believe that organizational longevity entitles them to exclusive privilege. They are prone to narcissism with an inflated sense of self-worth and entitlement. We have seen this toxic quality manifest in several behaviors and put into several general buckets.

    a) I have paid my dues—Someone did a good job last week, last month, or last year, so now they rationalize that it is okay to let up. In effect, this person is saying, "I don't need to work as hard now." Contrast that to good companies, where there is an atmosphere of teamwork and where people understand that dues must be paid every day if the team is to compete.

    b) I have a lot of experience here—Experience alone is not a tangible commodity. A history of showing up at the workplace in itself doesn't merit special reverence. While experience may be important, unless it delivers better performance and enables higher skills, then it is without value. In effect, the people who say this are really saying, "I don't have to be part of the team. I don't need to justify my actions." Having experience should never trump hard work and execution. At one point or another in a career, most of us have encountered a Rebel like this. We recall one job where a longtime manager was heard to say that he was a resource for those of us with less experience in the company. Immersed in his self-importance, he spent his days walking about the facility or sitting in his office, waiting for someone to pay homage and seek out his wisdom. He did nothing daily to help advance the mission. Everyone in the department resented the fact that upper management didn't require him to deliver like the rest of us.

c) It's not my job—Saying "It's not my job" is akin to saying, "I don't care about the team ....I only care about me." This attitude destroys team performance and can quickly turn a unified team into a troubled collection of individuals. Good companies establish teams that demonstrate willingness to do whatever it takes to be successful, as long as it isn't unethical or illegal. In fact, good people look for problems and intervene before being asked.

3. <u>You make the rest of us look bad</u>—Peer pressure can be a persuasive force, be it good or bad. Unfortunately, we have seen a lot of bad peer pressure applied to stifle good performers. This phenomenon seems to be one of human nature's vexing qualities. Putting down high achievers starts as children. Think about the girl or boy in your school class who always had the answers, turned in completed assignments, and aced every test. These kids were typically scorned and isolated by other children. Sadly, as adults, this trait remains fully entrenched and, in some instances, strengthens. Albert Einstein nailed it when he said, "Great spirits have always encountered violent opposition from mediocre minds."

Good-performing employees are usually self-starters. They are motivated and they compete most highly against themselves. They want to do better today than yesterday. These people attack every pursuit in the same fashion, whether it is doing a project at work, washing the car, or writing a letter.

On the other hand, the weak employee tries to skate and only do enough to get by. There is no want to win, only not to look bad in the shadow of a high performer. As a result, this Terrorist will pull out the stops to demean the high performers through gossip, sabotage, or ridicule.

4. <u>Gossip crusader</u>—Gossip crusaders have one singular mission: spread rumors and discontent. These people thrive on drama and disruption. They have little priority except to talk about other people. In turn, they waste time better spent on positive tasks for the customer while actively diminishing the dignity of coworkers. We have heard people say that gossip is human nature and it does little real harm. We disagree. Gossip is like a lit match. On the surface, it is just a tiny flame that seems harmless. But left unattended in the wrong time and place, that tiny flame can lead to the destruction of a forest. Good leaders don't allow a culture of gossip to prevail.

5. <u>Glory keepers</u>—Nothing important is ever accomplished alone. Even if a star player did most of the work, good and sustained results inevitably require a team effort. People who say, "It was all my idea" or "I made it happen" is like saying "I am more important than everyone else on the team." This attitude causes other members to harbor resentment. That leads to loss of team amiability and lessened productivity.

6. <u>Under the bus they go</u>—Along with "glory keepers," these Rebels are masters at avoiding accountability. No matter what has happened, it's always someone else's fault. The only time these people want to accept responsibility is if there is a positive result. In a team setting few acts are more selfish than saying "It wasn't me," especially when, at least in part, it was. In contrast, in good companies, teams have a view that win, lose, or draw, we are all in this together.

Focus on the Good Ones and Eliminate the Rebels and the Terrorists

Many of the Rebels and the Terrorists will eventually self-identify. Most are not clever enough to be totally clandestine due to large egos and a sense of invulnerability. But both types need to go—the sooner the better. They create havoc, and the rest of the organization is watching. To tolerate them sends a very damaging message. That is why it has been puzzling to us why leaders spend so much time and effort trying to motivate the poor performers when ultimate success depends upon those who are on board or on the fence awaiting encouragement. It is an affront to the Good People to tolerate the Terrorists and Rebels. As someone once said, "Remove from your ranks those who don't measure up. Do it as charitably as possible, but do it with a ruthless focus on the needs of your most talented people."

This is a key takeaway—people are the key to any successful change, and how they respond to the CEO's vision will make or break the transformational effort. In chapter 2 we emphasized that good companies know that attracting, training, and developing their people is their ultimate competitive advantage. They develop quality processes to recruit and retain people with better skill sets, attitudes, and cultural fit. This brings to good companies people motivated to succeed and willing to give differential effort to benefit customers and safely get the mission accomplished. Bad companies don't get any of this and therefore consistently underperform.

Transforming organizations, in substance, is about transforming people and convincing them that the CEO's vision benefits them. There is pride in belonging and belief that what they do is beneficial and meaningful. Engaging, aligning, and empowering a workforce is one of the most powerful things a CEO can do to win in the marketplace.

Stay the Course—Even When It Hurts

Consistency in values and message will define the sustainability of your culture. To employees, the CEO is the owner of company culture. You are on display, and your people will do what they see, not necessarily what they hear. That is why it is important to stay the course and focus on taking care of issues that challenge the culture—not only when things are going right but particularly when the team is task-saturated and prone to distraction. During such periods it may seem more expedient to let things go and hope for the best. But letting standards slip is dangerous, and resulting cultural wavering tends to destroy morale. Being consistent, no matter the challenge at hand, goes a long way to anchor the right culture with employees and stakeholders.

There are many examples where we see leaders conceding their values and jeopardizing team loyalty. One of the most visible examples in our society today occurs in athletics. The pressure to win may be so unyielding that deviant behaviors by key players may be overlooked despite conflicting with team values and team rules. For example, say that a star player violates team rules by missing a meeting, or openly criticizing the coaches in public because he feels slighted with reduced playing time. Rather than standing firm on basic values, because he may be a key player on the field, the coaching staff and administration look the other way. By letting standards slip, they have sent a strong and clear message to everyone else that values are to be observed only when it benefits the circumstance. This guarantees loss of team essence and demotivation of the good people who follow the rules and put team first.

Leaders need to ensure that teams remain focused on the assigned targets. They can't let their guard down, nor can they get frustrated. Sometimes progress is slow and there will be setbacks, with naysayers trying to claim defeat at every bump in the road to change.

Good Leadership Habits That Ensure Team Effectiveness

There are a number of positive leadership habits that we have developed over decades in organizational life. None of these habits are new, but we believe all are worth consideration as part of the equation to keep a team healthy and focused.

1. Never assume that things are fine. Business life is dynamic and fluid. There will be change, and nothing stays the same. The Greek philosopher Heraclitus (c.535–c.475 BC)

was alleged to have said, "No man ever steps in the same river twice." Whether the change is good or bad is up to you. There is no room for complacency. Every day counts in nourishing solid team character. Good leaders constantly assess team performance and look for gaps.

2. Never take good employees for granted and never assume that they will always be there.

3. Never compromise on baseline values because team cohesion may be at stake if there is inconsistent messaging or actions.

4. Recognize that the organization will talk to you, so listen and watch for signals that reveal organizational climate. Reality is your friend. Pay attention and be quick to react if you sense that the culture is getting lazy. Watch for Rebels or Terrorists who may surface.

5. Have a plan. Good processes and team alignment will not happen by winging it. Assess and reassess the plan continuously and leave nothing to chance. Pay attention to details and insist on flawless execution of the basics.

6. Personally engage. Commit the time necessary to ensure that there is clarity around team roles and deliverables. Trust your people to do the job and leverage collective team energy to fill possible organizational gaps.

7. Team alignment is never permanently fixed. It is always under repair. Past success does not guarantee future success. Understand the importance of instituting the right team architecture. Regularly evaluate your people—their attitudes, capabilities, and suitability. You need the best people in the right place to maximize financial performance. Force-ranking people is a healthy exercise. The process alone provides a number of positive things for a

company. For one, ranking people ensures that you put the right people in the right seats. It also forces fence-sitters to get engaged or risk elimination, as well as helps you identify possible future management stars.

8. We know that teams function best when there are clear, consistent, and inviolable values. Team effectiveness is anchored by leadership engagement and policies that recognize and reward team players. Collaboration and innovation are trademarks of good teams.

Sustaining a Well-Functioning Team

In this chapter we made the point that building and maintaining a good team is not a destination but a journey worth taking. Well-functioning teams are clearly a competitive advantage in a highly aggressive marketplace. The fate of the organization may rely on how well the team functions, and how well the team functions is a direct outcome of leadership bearing. In his article "How the best company to work for works" (*Management Issues* magazine, November 16, 2016), Rod Collins wrote, "While low levels of worker engagement and poor corporate cultures might have been acceptable—albeit not desirable—in the recent past, organizations that continue to tolerate a dismal place to work may be creating their worst competitive disadvantage."

Brian Fielkow, CEO of Jetco Delivery, author, and recognized leadership coach talks about the significance of the "Three T's" as a core element in building employee engagement: Treatment, Transparency, and Trust. In substance, Fielkow points out that employees respond with higher levels of engagement, productivity, and professional pride when they are treated with respect, when managers are transparent and open with them, and when they feel trusted to do the right thing.

As we wrote in chapter 6, the CEO and his team must be ever mindful that the culture is always under repair. Leading change is a continuous effort. Taking your eye off the ball is a risky thing to do since there will always be forces at work to undermine or deter the effort. A positive team ethic is key to transformation—it is not a nice-to-have thing; it is a must-have thing. How the team works together, what they produce, and how they engage will determine the future.

Final thought: The team is the key to any transformation. Teams function best when leaders engage in dialogue, figure out what their people need and want, and evaluate what is doable and communicate regularly. Building successful teams takes leadership patience and forthright action. It is empowering the good people to buy into the vision, take accountability for performance, and passion to move the culture. Good leaders turn the tables—they isolate the naysayers and mobilize the good guys. They leave no doubt that the status quo is no longer acceptable.

Chapter 8

Building a Success Culture

"The best companies and the best leaders
understand the real drivers of business success: a
long-term perspective which focuses on customer
and employee relationships as the sources of
competitive advantage and an emphasis on
values and ethics as guides to decision making."

—Jeffrey Pfeiffer, *Harvard Business Review* (2011)

Jim: In many of the chapters in this book, we examine fundamental traits necessary to build and support a success culture. Building upon many of these qualities, in this chapter we offer expanded thoughts on several themes to add emphasis. At the outset, we acknowledge that there are many definitions of the term *organizational culture*. For our purposes, in the context of this book, organizational culture is the shared norms, attitudes, beliefs, customs, and self-image that structure how people interact to get things done within the group.

In practical terms, we like to think of culture as convergence of people and process at the working face of an organization. In well-functioning companies, people and processes are allied and coordinated. In dysfunctional companies, people and processes are generally random, unconnected, and even conflicting. One thing we learned early on is that if people and processes are not aligned, then there are major cultural and performance problems.

Research and experience have established that culture can be the ultimate differentiator in any enterprise. Companies with a success culture have a competitive advantage. Engaged people vested in the organizational success is the hardest thing for a competitor to replicate. On the other hand, a poor culture most certainly undermines performance, marginalizes financial returns, and jeopardizes long-term viability. Herb Kelleher, founder of Southwest Airlines, arguably one of the most impactful business leaders of his time, was quoted as saying, "It's the intangibles that are the hardest things for a competitor to imitate. You can get an airplane. You can get ticket counter space, you can get baggage conveyors. But it is our esprit de corps—the culture—the spirit—that is truly our most valuable competitive asset."

The Bricks and Mortar

The parameters of a success culture must be defined. Without a plan, it is highly unlikely that it will naturally evolve. So how do leaders promote and sustain people-process convergence and a culture of success? We found that there are some basic questions that must be asked and answered at the outset in order to progress a culture of success.

1. Define organizational success. What does success look like in your company?

2. Define expectations. What do we expect of our people in how they act, work, and create value?

3. Do we have the right people to execute and anchor the culture?

4. What metrics are best to guide the process? Remember to keep the metrics simple and relatable to everyone in the lineup.

5. Tackle the backroom processes. Make sure processes line up with the desired behaviors.

In our experience, there are also a number of elements that must be present to maximize efforts in generating a positive culture ethic in companies, including:

Recognize That This Is a Journey

As we emphasized in chapter 7, instilling a success culture is not a destination. It is not something that you achieve then go on to the next challenge. Building and sustaining a success culture is a journey that never ends. It requires nurturing with continuous leadership focus. The sixteenth-century Theologian Martin Luther, referring to the Reformation, said, "We are not yet what we shall be, but we are growing toward it. The process is not yet finished but it is going on. This is not the end, but it is the road." So it is with the journey to a success culture. As a leader, once you think your company is there, you most certainly are setting yourself up for dysfunctional creep that can make your culture sick. The real benefit is that it is the journey itself that will drive success. Striving to get better every day provides energy that powers improvement.

Create and Implement a Clear Organizational Vision

Back in the mid-1970s, during a military training exercise, a young captain was leading a group of a hundred or so special forces troops (the good guys) on an extended navigation exercise from point A to point C, through checkpoint B. The context of this specific drill was that the troops were behind enemy lines, so a straight overland march was not practical due to simulated enemy patrols in the area. Using the standard-issue military Lensatic compass and topographical maps, the goal was for the captain to safely lead the good guys through the territory without detection, using evasive tactics. However, when the good guys failed to arrive at checkpoint B at the appointed time, the military training cadre broke radio silence and called the captain via emergency radio inquiring where he and his group were located. "Are you lost?" was the question. The response from the captain in the field was classic: "We're not lost. We just don't know where we are."

Our experience is that people in dysfunctional organizations don't have a clue where they are or where they are going. They may have a *map* (i.e., rules, policies, practices), but without a destination, maps won't help them get to where they need to be. They, like the captain, are lost. Dysfunctional managers spawn dysfunctional companies. They have no vision, nor do they have an idea where to even begin the journey. As a result, like in our preceding story, the troops are subject to the aimless wanderings of a manager wholly uncertain how to find a way home.

In contrast, one of the first things a good leader does to build and sustain a success culture is to present a clear vision for the organization. This vision is what the company aspires to be. It defines the ultimate organizational benchmark. The destination. It is the North Star that guides daily tasks so that everyone in the group is clear on how their work advances the journey.

The vision must be easily conveyed and clearly understood at every level of the organization. It must be easily articulated from entry level to boardroom. Perhaps one of the most historic vision statements ever delivered was in 1961, when President John F. Kennedy challenged NASA to "Put a man on the moon and return him safely to Earth before the end of the decade."

More recently, one of the clearest vision statements we have experienced was "zero safety failures" in a major transportation company. When the concept of *zero* was introduced, it was thought an entirely unreasonable goal by most everyone in the company. "Too many things we can't control" and "As long as you have people involved, we will never get to zero" were common sentiments expressed. Through benchmarking world-class endeavors, the sheer resolve of leadership that *zero* was the only acceptable outcome, and the personalization of the impact of safety failures on individuals and families (e.g., it was asked "If zero isn't doable, then who here will volunteer to be one of those killed or injured on the job?), the point was well-made and the concept of zero was gradually internalized. With this mind-set, in a little more than five years, the company reduced its safety incident rate by more than 80 percent, and more than 75 percent of company operations actually achieved zero in locations where employees and managers embraced the approach. The real benefit in shooting for zero in safety lies in the journey. Striving every day to achieve zero failures brings quality processes that advance safety.

We have seen in some settings that conceptual confusion may develop in management circles about the difference between goals, mission, and vision. The way we keep it straight is to think about it like this:

- The Goals: The specific actions defined and measured at a given point in time. These actions support and enable mission accomplishment.

- The Mission: The mission is the pathway to achieve the vision. It is who we are today.

- The Vision: The vision looks beyond tomorrow. It is where we are going as a group. It provides everyone a line of sight to the endgame. It gets and keeps everyone on the same path going forward[5].

Crafting a clear organizational vision requires artful involvement of the team. Sitting alone in the corner office and constructing a vision may be expedient, but unless you gain consensus of the team, there is a good chance that people won't buy in and fully engage.

Once crafted, the vision has to be implemented. Legendary business guru Jack Welch said it well: "Good business leaders create a vision, articulate the vision, passionately own the vision, and relentlessly drive it to completion." That means that leaders own the vision and should be able to wire every business decision to vision attainment. Good leaders continually challenge their

---

[5] Avoid the trap of trying to incorporate in the vision statement the tactics to achieve the vision. The "how" is best defined in strategic plans and individual execution tactics.

teams and ask the question: If it doesn't support the vision, why are we doing it?

Craft Foundational Values

The walls, framing, flooring, doors, and roof are all critical components when constructing a building. But nobody would try to build a structure without a solid, level foundation upon which to securely affix the walls. In organizational terms, the vision, mission, goals, resources, and strategies are important components in a company's structure. And like the walls of a building, unless there is a solid foundation upon which to build, these components won't stand. While a house structure is founded in concrete, successful cultures in business must be founded in a set of solid values. Your corporate values define who you are. They are an organization's DNA—the air in the corporate lungs, the blood in the corporate veins. Everything else in the organization builds upon the values.

There is power in value-driven leadership. Values link everyone in the company and ensure consistent, well-defined behaviors are the norm. They help ensure that you will stay on dry ground with consistently ethical behaviors despite inevitable challenges and contradictions in the marketplace. Values are your cultural guardrails that keep everyone on the right road in pursuit of the vision.

Long-term corporate success depends upon the ability of everyone in the company to execute with integrity every day. Actions and words frame organizational reputation in the marketplace and shape a company's ability to compete. Customers, shareholders, communities, and regulators count upon companies following the highest standards of business ethics.

Leaders are the example in living the values, and through their example they demonstrate an expectation that people in their charge will conduct company business according to organizational values.

There are many examples of corporate value presentations. In our opinion one of the most effective is Patriot Rail Company's value proposition. Under the leadership of CEO John E. Fenton, the company adopted a unique structure to present their values by putting them into three interactive buckets as follows:

How we work
>> Safety—our DNA

>> Caring for our environment and our coworkers

>> Focusing on our customers and providing solutions to help them succeed

How we act
>> With financial, ethical, and personal integrity

>> As a team—collaborative and with respect for each other

>> Communicating openly and honestly

How we create value
>> Engage our employees

>> Optimize our financial, physical, and intellectual resources

>> Simplify, measure, and evaluate our processes regularly

These values are communicated regularly, from the top down through the organization, and are taught to every new employee and manager in formal orientation when hired. Examples of how to live these values are regularly discussed within the team.

Printing lofty values on a poster and hanging it in the lunch-room or hallways alone will not get you where you need to be. Values must be living, relatable attributes that employees know how to apply in their jobs. They must see the example of their leaders living the values in every decision.

Establish a Just Culture

A lot has been written about the term *just culture*. Many academics and learned people have offered definitions of the term. As practitioners, what we know is that without just culture, there is little opportunity for a company to build or advance a success culture.

In substance, creating a just culture is really all about the Golden Rule ethic. Just culture is a management system where people are treated with fairness, consistency, and transparency. Just culture is a balanced business standard with attributes that provide workplace dignity, justice, and accountability. Everyone plays by the same rules and there is equity in treatment. There is no preferentialism or special favor. There is a "heaven" with open recognition for those on board and aligned; and, there is a "hell" with definite consequences for those who try to subvert the culture—be it passive or aggressive in nature.

A just culture is a system that also establishes an employee-centric approach to customer service, in lieu of putting the customer as the principal focus. With this approach, the benefit is an improved customer experience as a by-product of a highly engaged, aligned, and positive workforce.

Just culture methodology also assigns responsibility for performance and execution to the organization, as well as the individual. This collective-cultural standard is designed to help boost organizational performance by placing less focus on incidents, errors, and outcomes, and more focus on risk identification, system design, and the management of the behavioral choices of its people. This, in turn, makes employees feel more valued as business partners vested in the success of the brand. Richard Branson was credited with saying, "If you take care of your employees, your employees will take care of your customers and your customers will take care of your shareholders."

Alignment—Make It Personal

> "Coming together is a beginning. Keeping together
> is progress. Working together is success."
>
> —Henry Ford

Once a clear vision and detailed values are established, gaining team alignment is the next step in the journey to build a successful culture. Developing and maintaining good alignment is the baseline catalyst for performance, profitability, and growth.

Some Alignment Basics

- The first step is to ensure that the leadership team is aligned. An organization, to paraphrase Ralph Waldo Emerson, is the lengthening shadow of its senior team. When the senior team is not aligned, the organization cannot be united in purpose.

- Legend, lore, and custom. Over time, every organization develops unofficial ways to do things. These informal ways may end up as a practiced custom and may even conflict with established protocols. In order to achieve and maintain organizational alignment, management must ensure that what is seen and done is the same as what is said and written. Whatever the form or content, legend, lore, and custom must support the desired cultural norms and expectations. If they don't, you'll find proliferation of silos and an at-odds atmosphere between and among departments.

- Alignment is a process that is ongoing. Business and life are dynamic and things change. Maintaining alignment must be a continuous leadership focus.

- Getting aligned and staying aligned isn't easy. It is hard work, and it can be messy. But maintaining alignment separates the great organizations from the rest.

- Gaining alignment does not mean that everyone agrees 100 percent with what needs to be done. But it does mean that people have the opportunity to participate in policy discussions and have their points heard. At the end of the process, however, alignment requires that everyone buy in and engage in the decision that was reached by the group, even if they disagree.

There are some specific methodologies that help facilitate organizational alignment and better governance. We have worked with Brimstone Consulting Group (Brimstone) on a number of projects over the years. The Brimstone team developed what we believe is a highly effective methodology to guide leaders in efforts to align their organizations. Brimstone's foundational premise is that good organizational alignment not only accelerates decision-making; it also enhances employee engagement and improves customer satisfaction, which leads to elevated performance and profitability. Brimstone uses a great metaphor to tee up the importance of good organizational alignment. They say that, in a sense, an organization is like a motor vehicle. In a motor vehicle, in order for the driver to achieve maximum performance, all systems must be in alignment. When a vehicle falls out of alignment, it is typically evident through feedback to the driver who can "feel the road." However, as Brimstone points out, misalignment of an enterprise may not be as quickly apparent as a malfunctioning vehicle. Organizations can be complex and senior leaders can be isolated from real feedback (i.e., they can't "feel the road"). Brimstone has developed a seven-step cycle to promote alignment in organizations of any size:

1) Build a common understanding of market realities and impact on the business. Unless everyone agrees on the risks and realities facing the group, there can be no alignment.

2) Engage in dialogue. Talk about the real issues in candid terms. Be honest and accept the reality.

3) Manage internal conflict. Candor invariably generates conflict. But healthy tension is not a bad thing in organizations, if properly managed and resolved in constructive ways.

4) Reach agreement. Once everyone has a chance to offer opinions, the group must make a decision. As touched on in No. 3, not everyone has to be happy with the decision. But all stakeholders should have input into the decision, and they must support the decision once made.

5) Clarify decisions. Everyone must understand what has been decided and what needs to happen next and by whom. Ensure everyone is on the same page on next steps.

6) Create ownership. Everyone must feel like they have a stake in the outcomes. Communicate and circulate the rationale and the need for the action.

7) Take collective action. Once decided, everyone on the senior team marches forward and supports the action and executes the plan.

Leadership

> "It appears that the new metric of corporate leadership will be closer to this: The extent to which executives create organizations that are economically, ethically, and socially sustainable."
>
> —James O'Toole /Warren Bennis, "A Culture of Candor," *Harvard Business Review* (2009)

There are thousands of books that have been written about qualities of leadership. We won't rehash all the principles here, but for our purposes in this chapter, we want to emphasize several specific central elements of good leadership that help drive a success culture.

Leadership Trust

General lack of trust in leaders and establishments is a major issue in our society today. Many traditional institutions are losing the trust of the public they serve. In the context of business, Steven Covey reported that research shows that, in general, only 49 percent of employees trust senior management, and only 28 percent believe CEOs are a credible source of information.

Stating the obvious, then, unless people trust their leaders, there can be little hope of high engagement. But as Covey points out in his article, trust is only part of the equation. People must also believe that their leaders are competent and understand the realities. Covey says, "With the increasing focus on ethics in our society, the character side of trust is fast becoming the price of entry in the new global economy. However, the differentiating, and often ignored, side of trust—competence—is equally essential. You might think a person is sincere, even honest, but you won't trust that person fully if he or she doesn't get results. And the opposite is true. A person might have great skills and talents and a good track record, but if he or she is not honest, you're not going to trust that person either." People grant trust to leaders only when that trust is earned. Earning trust requires leaders who engage with their people, listen to their concerns, take action when warranted, and communicate openly.

The Privilege and Responsibility of Leadership

We believe that leading people in business and life is a privilege. It is a higher calling and a responsibility that cannot be taken lightly. Leaders make decisions that affect the lives and careers of employees and their families. They make decisions that can affect customers and communities and any number of other stakeholders.

With the privilege comes responsibility. Leaders are on a pedestal. Their movements are on display and they are watched closely for signals. "The people follow the example of those above them" is a Chinese proverb that goes to the core of this aspect of leadership. Leaders can't underestimate the impact of their words and actions.

Another aspect of leadership responsibility is the imperative to support people in their work. Those in the ranks are responsible for doing their jobs well. A leader, on the other hand, has the responsibility to support workers individually and collectively. This support could mean committing additional resources, or eliminating burdensome bureaucratic red tape that inhibits efficiency, or adjusting priorities to add clarity to deliverables.

Communicate and Stay Engaged

In chapters 2, 4, and 6 we examine at length the importance of CEO engagement and communication with all levels of the organization. One of our key points is worth repeating here: Leadership can't be delegated. Good leaders directly engage and stay connected with people in their organizations. They develop an honest, ongoing dialogue to help them stay connected with the reality of life in the trenches. This is an important issue because, typically, the higher one goes in management, the more isolated from real happenings they become.

Another plus with meaningful direct senior-level communication is that senior management can control the tone and ensure key messages are not watered down or altered when allowed to filter down through organizational levels.

Be Willing to Lead Loudly

In chapter 2 we wrote that the inability of leadership to connect with people is a common trait in cultures of failure. Leaders who are detached or insincere are incapable of driving the kind of change that moves people to new levels of performance. Bob Marcus, founding partner of Nvolv Partners Inc. and longtime leadership coach, put it well: "After working with executive leaders for more than two decades, one of the things I've come to believe is that in times of change or crisis, leaders need to be prepared to lead loudly. Messaging needs to be consistent, cogent, and constant, and a leader's actions big enough for the audience in the back row to see. In the pressure to deliver results every day, many leaders forget that a big part of their job is to model the behaviors they want to see in the organization."

Know the Games People Play

Successful leaders understand that people view the world through different filters based upon life experiences and outlook. Everyone has their own cultural filter through which they view life, events, interactions, and relationships. We have seen how self-interest drives human behavior on the job. People want to know how they will be affected by initiatives or projects. Good leaders message initiatives and policies in a manner that appeals to personal interest of the people in the organization. Creating a compelling case for change accelerates buy-in. People more readily adopt and internalize the mission at hand. They are aligned and ready to do their part.

Perhaps one of the most striking relatively recent examples of using the self-interest phenomenon to align a skeptical team is Gordon Bethune. We briefly touched upon Bethune earlier. To expand, in 1992 Bethune, a former aircraft mechanic, pilot, and Boeing executive, was hired to lead troubled Continental Airlines. As pointed out in the article "Up in the Air" in the November 2015 *Texas Monthly*, author Loren Steffy wrote that, at that time, Continental was the worst-performing U.S. airline in all four service benchmarks measured by the U.S. Department of Transportation: 1) on-time arrivals, 2) lost baggage, 3) customer complaints, and 4) involuntary denied boardings. To boot, the company had not posted a profit outside of bankruptcy since deregulation in 1978.

To improve customer service and improve performance, Steffy writes that Bethune and Greg Brenneman (recruited by Bethune as Continental's new chief operating officer) knew that first they had to boost employee morale. "They made the employees very much a part of making Continental a better airline [and] combined the feel-good aspects of their turnaround with careful scrutiny of costs and created incentives that, this time, aligned employees' interests with those of passengers." As Steffy reports, instead of offering a bonus for pilots for fuel savings (instituted as incentive to improve on-time performance before Bethune took control), they rewarded every employee with a $65 check every month that the carrier finished in the top five airlines for on-time performance. They also implemented a profit-sharing plan for employees. While $65 may seem a paltry sum, Bethune, in his typically forthright way, said to the effect that "even a millionaire will bend down and pick up three twenties and a five-dollar-bill if it is on the street in front of him." The $65 monthly bonus was provided in a separate check, not included in the payroll check. This was to leverage the optics around it being a special incentive.

The results of Bethune's innovations were soon apparent with the airline making a profit of $556 million in 1996, while the stock price soared from $3.25 to $50.00 a share. Continental quickly moved up in the performance in every category of the U.S. DOT ranking. Steffy writes that "The turnaround was so impressive that [other] ailing companies ....started poaching Continental's management talent."

How It All Works

In summary, a success culture requires a healthy team ethic, as outlined in chapter 7. People want to be successful and they want to feel like they are on a high-powered team, doing meaningful work and making a difference. We have all heard the axiom that a great strategy with a bad team will struggle but a good team can make even a bad strategy work. For example, if the military captain in the story above had been a good leader with a good team, with the right resources, a good plan, and disciplined execution, they would never have gotten lost, because the captain would have been able to keep his team on course. Instilling and anchoring a success culture is a journey worth taking. The journey requires shared vision, clear values, team alignment, a just culture, leadership engagement, individual accountability, and the right metrics to guide decisions and allocation of resources. We go into detail on these topics and other essentials in the chapters that follow.

SECTION III

THE BASIC BUILDING BLOCKS

Chapter 9

Key Initiatives

The Acid Test for Every Company

Chuck: In the second book section, "Overcoming Dysfunction," we talked in chapter 4 about the change process and the basics of fixing dysfunction in chapter 5. We went into detail about the CEO's role in chapter 6 and talked about building the team in chapter 7 and transforming the culture in chapter 8. Now that the infrastructure for change has been addressed, we are ready for the third section of the book, "The Basic Building Blocks." In chapter 9, it is time to expand on the change process in chapter 4 by getting into the big engine for change within a company, the key initiative. We need to understand this subject in detail before we get into the key changes in safety, HR, and performance management presented in chapters 10 through 12.

The actual transformation of a dysfunctional company into a good company or the upgrading of a good company into a world-class company involves key initiatives. In fact, many of the chapters in this book are devoted to key initiatives involving safety, people, and performance management. Anytime a company wants to take a "great leap forward," the key initiative is usually seen as the vehicle to do just that. The Six Sigma and Work Out programs adopted at GE were key initiatives that not only made GE more effective and profitable but helped define or at least redefine the GE culture not only for those on the inside but for anyone watching that company.

All kinds of companies from the world-class level to the most dysfunctional try to undertake key initiatives, because they can be a powerful means of unifying the organization in a focused effort to dramatically transform who the company is, how it operates, and what it can accomplish. However, there is a more sinister side to the key initiative, and that is that it very clearly and rather quickly tells you who you are. If you have the right leadership and the right workforce, and you have picked the right initiative and done all the necessary groundwork, then there is a good chance that the key initiative will meet with success provided everyone involved does their part. But if one or more of those necessary ingredients is missing, then there is nothing but failure waiting for you in the near future. And that failure is expensive and absolute.

The world-class companies generally have the infrastructure to successfully undertake a key initiative. Most of them have a long track record of making key initiatives stick in the company. If there is a failure, it usually ends up as a case history in many MBA programs because it is a surprise to practically everyone, the impacts on the company and those leading the effort are usually quite severe, and all the "Monday morning quarterbacking" that follows produces a host of reasons for the failure. Good companies can also be successful with key initiatives, but they have much less in the way of tools than the world-class company, so they must be more careful. The scope should be more modest, and the level of effort needed from the leadership is much greater because the team has more holes and is more prone to making mistakes. Plus, the history of success with key initiatives is probably not there. This will be a *maiden voyage* for many of the participants, and that can have a huge adverse impact on the probability of success, especially when things start to go wrong. Needless to say, the dysfunctional company has virtually no chance at success unless the decision has been made in advance, independent of the key initiative, to transform the company. Unfortunately, that is usually not the case. In fact, in many cases the dysfunctional company sees the key initiative as a miraculous way to change the company for the better—without anything else being required. That is a fallacy that has brought many key initiatives down.

It should be apparent that while the key initiative is a powerful tool in the arsenal of any company, it has to be employed with great care because failure is often a more common result than success. We will talk about the reasons for those failures in the section "Why Key Initiatives Fail" and then use those lessons to construct a proper key initiative program later in the section "How to Do it Right." But first, we need to define exactly what a key initiative is, because a poor understanding of the key initiative is often a major cause of the failure that results.

What Is a Key Initiative?

Very simply, a key initiative is any change process that affects a large cross-section of the company in a very significant way. It can be a total change in how information is collected and utilized, how goods and services are procured, how customers are served, or how employees are managed. Key initiatives can make companies more productive, more creative, more responsive, better managed, safer, and more nimble. The scale of change can be huge. They are a big deal and not just flavor of the month. The earlier references to Six Sigma and Work Out at GE are perfect examples of key initiatives. These two programs changed forever how GE workers approached their work and how they managed what they did. Everyone in the company knew about these programs and was touched by them in a number of ways. People had to learn new skills, and new terms were added to the company lexicon. Conversations in meetings changed and the makeup of teams on new projects now had different members who had new skill sets and duties. These programs changed the company for the better, and basically everyone in the company and outside the company knew about them. That is why key initiatives are so important. When they work, they can have a massive positive impact on the company.

Key initiatives are not new processes or changes in processes in an isolated part of the company that may make that part of the company better but have no direct impact on anyone other than those people in that part of the company. These types of process projects are very important and they are covered in a later chapter, but calling them a key initiative, managing them like a key initiative, and getting everyone involved in them like a key initiative is not warranted. That just leads to unneeded complexity and key initiative burnout, and the chances of that isolated project being successful just go down.

Key initiatives consume a lot of leadership time and energy as well as manpower in the rank and file. They can be very disruptive to the normal workplace and the final cost is always significant. So it is important to do them sparingly and only when there are excellent reasons and a great expected payoff. Since they are so expensive, in terms of real dollars as well as time and energy, it is critical that all the key elements be in place at the beginning so that the chances for success are maximized. This is a great opportunity to lead and manage, but if you are relying on hope to see you through, then you will be disappointed, as you will see in the next section on how key initiatives fail.

Why Key Initiatives Fail

A new CEO at a large publicly traded company that had issues wanted to make his mark early, so he hired a consulting firm—which suggested that he start some key initiatives. The CEO brought in one hundred of his top managers for a weeklong session at the corporate office. After two days of discussions, which were dominated by the CEO, the lead consultant, and probably half a dozen other people, six key initiatives were selected. The one hundred managers were broken up into six teams and spent the rest of the week developing the details of the initiatives, assigning roles, and determining schedules. Each initiative had an assigned consultant, a senior executive from corporate who functioned as a sponsor, a leader from the manager group, plus other managers who had been assigned specific tasks. At the end of the week, the CEO and the lead consultant both made speeches about how important these initiatives were to the company, and everyone left with high optimism. Over the next three months, progress was made on all six initiatives and there were additional meetings and update reports to the CEO. And then slowly over the next six months, they all ran out of steam. Some of the initiatives encountered problems that made them more difficult and progress virtually stopped. In some cases, people argued over the details, and consensus could not be reached about what to do next. People got busy doing other things and consultants complained about the lack of cooperation. Eventually, the CEO stopped asking for progress reports, meetings that were postponed were never rescheduled, and the six initiatives just disappeared. So what happened?

This new CEO and his consultant made just about every mistake that can be made in getting a key initiative off the ground and running. First off, the company had a history of dysfunction and failure. The previous CEO, who had just retired, had been brought in five years previously to rescue the company from near bankruptcy caused by myriad bad behaviors at all levels within the company. That turnaround CEO had initially "taken on" the Rebels and the Terrorists who controlled the company and had made some modest progress in changing parts of the company culture and building some processes in a company that operated without processes. However, although he had gotten rid of a number of the bad actors, there were still a lot of the Rebels and Terrorists at high levels in the company and they had begun to take back the company in the previous twenty-four months when the turnaround CEO started to lose interest. The new CEO was not a seasoned executive and was no match for executive-level Terrorists who had been manipulating CEOs for much of their careers. In fact, he soon adopted their view that all operational decisions should be made in the field and that corporate involvement in virtually anything was not warranted. Almost immediately, the programs built by the previous CEO began to unravel and were virtually gone in the following two years. At the time the six new key initiatives were started, the company was well on its way back to where it had been culturally five years previously and was in no shape to take on any kind of universal improvement effort, let alone six of them at the same time. In fact, killing these six key initiatives was a good way for the lead Terrorists to show the rest of the company that they were back in charge. The first requirement for successful key initiatives is that the company environment, which includes the leadership at all levels and the culture throughout the company, must be ready to take on such a significant project. In this particular case, nothing could be further from the truth.

The second requirement is that the CEO has to create a sense of urgency and get people aligned behind his vision of where the company should go. As we have previously pointed out, John Kotter, in his book *Leading Change*, writes that there must be a sense of urgency throughout the company before any kind of a change process can take root. The CEO must build a case for change that every rational person can understand and get behind. For real change to happen, it cannot be business as usual. People in the organization have to believe that the status quo is tantamount to failure. The company must change and everyone must get behind that change and do their part. In this particular case, there was no sense of urgency. The CEO never talked about how the company had to change and how these initiatives were going to take the company to a new level. He never defined where he wanted to go and why that was important. He just wanted to do some key initiatives because he was told that he should. He had no vision that he could communicate to get people excited and moving in a common direction. The one hundred managers saw this overall effort as just something the new CEO wanted to do and felt that it would pass. They were not excited nor did they feel the company was at an important crossroads. They did the minimum to stay out of trouble, and most were probably relieved when all six key projects disappeared. These six key initiatives were not a big deal to anyone. In the end, the CEO never talked about the failure. The consultant got paid and moved on. The lead Terrorists gloated over their success, and everyone else just shrugged and went back to their normal jobs.

The third requirement is that the CEO must own the process and then build a coalition of powerful leaders who believe in the effort and have the authority and the ability to take it forward. Gordon Bethune—in his book *From Worst to First*, about the turnaround of Continental Airlines—says that you have to have a leader on point who inspires and that you have to engage personally and not just delegate responsibility for the change. John Kotter goes on to say in *Leading Change* that you have to form a powerful guiding coalition around that leader, and that group must have the necessary power to make the change happen. None of that took place in this case study on key initiatives. The CEO merely stated that he wanted to do some key initiatives and had little idea what they were, and then he handed all the work off to other people and stayed remote. He gave out work assignments, but he never communicated a vision that got people excited when they saw its potential; he never articulated why all of this was so important, and he never owned any of it. The people who actually took the work forward were not a coalition. They had responsibility (although it was fleeting), but they had little authority. When things went wrong, they had no power to get things back on track. And most importantly, they never really bought into the program themselves. They were just drafted to do the work and they moved away from the initiatives as soon as they could. The key initiatives were orphans from day one—and that preordained their demise.

The fourth requirement is that the program of key initiatives must be feasible given the overall capability of the company from its leadership to its rank-and-file employee base. Even the most capable world-class company has to be prudent and not try to "boil the ocean" with multiple change programs going on at once. It has been pointed out already in this chapter that key initiatives demand a lot in energy and resources. In most cases, more than one is too much, and a really complex initiative can be too much. The CEO has to be realistically confident that the required work can be done with quality and the results institutionalized into the organization. A company taking on complicated initiatives or multiple initiatives must have top-notch leadership, mid-level managers, and employees. The company should have a long track record of success with key initiatives, and there should be virtually no significant internal issues that could complicate the program and possibly bring it down. The CEO should be totally confident that the company can complete these projects in a timely manner while doing everything else it must do to make the year successful and meet targets. For the company in this case history, one key initiative was probably too much—let alone six. The fact that six initiatives were picked just demonstrates how clueless the CEO and the consultant were about the whole process. Change is naturally difficult for virtually everyone. Even top-level people can burn out on change, so it is crucial that the CEO be in touch with his people at all levels and understand their appetite for making big changes at any given time.

The fifth requirement is that the CEO has to stay involved and follow through. The book *Execution* states that leaders have to follow through on what they want done. A CEO may delegate much of the work, but he cannot delegate ownership. The second and third requirements basically state that the CEO must initiate the vision and energy for the key initiative project and then recruit a coalition to the effort that also has the necessary energy and can activate the vision. However, that does not let the CEO off the hook. Coalitions can wither and die in the face of opposition if the CEO steps out of the picture. CEOs have to stay involved, keep asking questions, praise successes, participate in the needed fixes and changes, and continuously beat the drum. It should be obvious to everyone inside the project and outside the project that the CEO is not going away, is not losing interest, and will not be distracted. Anyone who quits has to answer to the CEO. That is how projects of all kinds stay alive and ultimately find success. Key initiatives cannot make it without the CEO front and center. In this case history, the CEO was not front and center on day one and stepped further away with time. People watched the CEO quit, so it was easy for them to quit as well.

Another way that a key initiative can fail is at the end. The CEO and the coalition can move the project along and all the required activities can be successfully completed. The CEO can declare victory and life returns to normal for the coalition members. However, with no one looking, the functional managers never truly adopt the new behaviors. People say they are doing things in the new way, but they really are not doing them for a variety of reasons. Eventually, remnants of the key initiative may still be around in form only but are having no impact on how the company functions. John Kotter states that the new approaches that come from change processes must be institutionalized into the very fabric of the company. All managers must take ownership and ensure that their people adopt the new behaviors and master the skills needed to do work in the new way. This is another responsibility for the CEO, because he is the only one who can do it. For the CEO, the key initiative is never done. The CEO has to keep asking managers how their people are doing with the new way of doing things, get feedback from rank-and-file employees about what they think of the new way, and be open to input about improvements or needed changes. People need to know that the key initiative is still important to the CEO and that it better be important to them as well. In this case history, the key initiatives never got to the institutionalization phase, but it is a sure bet that had even one of them gotten that far, it would have failed in this final step.

This case history is admittedly an extreme one but it did take place, and others like it happen every day. Wanting to improve the company is a good thing, but knowing how to do it is crucial. Bad companies are always in a power struggle between the good people who want to see the company reach its potential and the bad people who want everything to stay the same. When initiatives fail, it is another blow to the good people who lose heart and make plans to go somewhere else, and it empowers the bad people who win another one and then tighten their grip on the dysfunctional company.

How to Do It Right

We are going to illustrate how to do it right with another case history. But first let us review the six necessary conditions for a successful key initiative that were presented in the previous section.

1. The company must have the infrastructure, including the right leadership and employee base, as well as the right company culture to take on a major change effort.

2. The CEO must create a sense of urgency and get people aligned behind his vision.

3. The CEO must own the change effort and then build a powerful coalition to take it forward.

4. The change effort must be feasible given the overall capability of the company.

5. The CEO must stay involved and follow through.

6. The change effort must be institutionalized into the fabric of the company.

Anyone who has done a successful key initiative could look at this list and suggest some additions. We could end up with a list containing fifty items. However, we believe these six conditions are the critical ones, because if you have these and you stay true to the change effort, all of the others will take place.

The case history we want to present involves a large service company with massive safety problems, in addition to about any other kind of problem you could imagine. The turnaround CEO who came in to clean up the mess had a lot to deal with, but he immediately saw that safety needed to be addressed early in the overall change effort. Like most prudent CEOs, he knew that companies cannot excel operationally if the safety program does not work. This is pointed out and expanded on in the next chapter on safety. The CEO asked one of his SVPs to take the lead on this project but made it clear that he would stay involved. The SVP did not have a safety background, but he knew dysfunction when he saw it and was determined to make the company a safe place to work. He dug into the safety data, what there was of it, and interviewed all the main safety "leaders" in the company. What he found was not pretty, but it told a very clear story. The company was routinely injuring more than 20 percent of its employees annually, and fatalities were well into double digits. The safety leadership was not aware of any of this until the SVP presented it to them. They were not outraged by any of this and felt that it was just the "cost of doing business." They all gave their programs a grade of A- to B. The Safety VP was no better and, basically, acted as an apologist for the program in general. There was an excuse for everything, and nothing could be fixed. This person's main priority was making sure everyone had a hard hat and a safety vest. Another disheartening observation was that none of management felt any ownership for safety. At any level, when the SVP asked a manager a safety question, the response was, "Talk to my safety person; I do not deal with that." Condition No. 1 states that the company has to have the infrastructure for change. At this point *only* the CEO and SVP were on board. A lot of people had to go, and the rest needed to hear a new message that they were going to be expected to embrace. It was not going to be pretty.

The next crucial step was hiring a new Safety VP. A recruiter was engaged, because outside talent was needed for the position. The right person was not available from within the company, plus hiring internally sent the message that the program just needed a tweak. That was the furthest thing from the truth. The SVP interviewed twenty-four people for the position. At the fifteenth person, the SVP almost settled on someone who was good—but probably not up to this effort. The CEO set him straight by saying, "I do not think he is the right person, but if you want him, then hire him. It is your call." The SVP went back to the drawing board, and with No. 24, the right person walked in the door. It was immediately obvious that this person had done it before and had the "fire in the belly" to take on a totally dysfunctional company. This is a great lesson on hiring people. You cannot win without the right people, so never settle. Keep digging until you get what you need. With this key hire, the SVP had what he needed at the top. It was time to start engaging the rank and file.

After the new VP got up to speed on the situation and was able to give his observations from a more learned safety prospective, the CEO and SVP introduced the new VP to the senior manager core. In this one event, the CEO communicated his sense of urgency. He said he was outraged by these safety results and would not tolerate them. He demanded change and talked about what he thought were acceptable safety results. He also made it clear that he was going to stay focused on safety as long as he was at the company and that the SVP and new VP had his unwavering support. After the SVP made his comments about what he had found, the changes he had made so far, and the impressive background of the new VP, he introduced the new VP. This gentleman started out strong by stating, "From this point forward, you do not delegate safety; everyone owns it." He talked about how world-class safety programs work and what he expected the company to accomplish: zero safety incidents. He also stated that he was not a quitter and would see this thing through. The people who wanted to get on board got the vision at that meeting. Those who did not knew they were in for the fight of their life.

The next step for the SVP and the VP was to build the coalition to create the plan and take it forward. Like all dysfunctional companies, the full cast of characters was present: Terrorists, Rebels, Good People, Acceptors, and Fence-Sitters. The SVP and VP talked to virtually all the top-level managers and a number of district-level managers about safety and what they thought of the need for change. They selected a core of managers at all levels who seemed to be engaged and wanted better safety. They also mixed in some Fence-Sitters and Acceptors and took a chance on a few Rebels who they thought might have potential. They made their share of mistakes on people and quickly addressed the resulting issues when they arose. Not all the problems came from the Rebels. A few Rebels actually contributed and brought in points of view that were new and surprisingly useful. Some Acceptors and Fence-Sitters turned out to be passive Rebels, and a few Good People were actually very practiced Terrorists. The SVP and VP (and the very capable staff that they hired) stayed attentive and fixed their mistakes as soon as they were identified. The rule they rigorously applied was that people could stay as long as their behavior was constructive and not obstructive. An effective plan and a set of rules and tools were built with much guidance from the VP and then put into practice. The initial effort was to have the staff work with districts where managers were the most receptive in order to build a *beachhead of success*. The SVP and VP focused on problem areas near the top, such as senior-level managers who were obstructive or functioning as Terrorists. There were many tense conversations, but it was clear that the SVP and VP were not backing down. The CEO got involved a few times and some people left the company. At times, it was so difficult and labor-intensive to make any significant gains that overstressing the company was a legitimate concern. But the team kept pushing and never wavered from its standards or from the vision. The team mantra was "Be a pain in the a** with class." On one occasion, the VP visited a particularly difficult operation with really bad numbers. The manager said he was mystified by the lack of progress, as they were doing the whole program. Then he took the VP for a site tour and did not buckle his seat belt, which was a major violation that could get a driver terminated. He soon got the lecture about leading by example. This person did not last much longer. A lot of the progress was connected to breakthroughs. One particularly difficult senior manager, with whom the team had experienced many heated conversations, finally saw the light and got on board. He was not a bad person, but he was set in his ways and he was stubborn. Once he was convinced, he had a huge impact and became a disciple for the program. Personal relationships and a lot of one-on-one dialogue (not monologue) were the keys to keeping the program moving forward. In this particularly difficult case history, following through at all levels was key to the ultimate success. The CEO has to follow through, which he did with regularity, but everyone else has to do the same.

One concern with all key initiatives is that people are just going through the motions and not really embracing the change. That is why institutionalizing the change so that it becomes part of the fabric of the company is so important. In this particular case, the company was very averse to metrics, which is typical of the classical dysfunctional company. The VP and his team developed a set of safety metrics that were fairly standard but were modified for the industry. These metrics were very useful and they became part of the lexicon of the safety team, but initially, you rarely heard them out in the field. The SVP participated in the quarterly business reviews and it became standard practice for him to ask managers what their safety metrics were, since they were not reporting them during their presentations. In virtually all cases, the response was that they did not know those numbers. The SVP would then counter by asking them to explain why they did not think those numbers were important to their business. After a number of embarrassing periods of silence, managers finally got the message. They not only had the numbers; they put them in their reports and made comments about issues related to the metrics. Eventually the metrics became part of the normal lexicon in most operations and managers began using the safety metrics to hold contests between their various operations groups. Robust discussions about safety results became a normal part of quarterly reviews.

With this elevated senior leadership visibility, along with instituting a mind-set that *zero* was the only acceptable goal, as highlighted in chapters 8 and 10, the company was eventually able to reduce its safety incident rate by more than 80 percent, but that only tells part of the story. Further analysis showed that more than 75 percent of the company operations were at zero, which meant they were safety incident–free. Those operations had safety programs that were operating at a world-class level. They had embraced the initiative from top to bottom and were thinking, operating, and communicating in a new way. However, the balance of the operations had virtually not changed. Their managers had decided to ignore the key initiative and the senior executives they reported to let them get away with it. The company was virtually split in two pieces, with one large group moving forward and getting better. The other, smaller group had not changed, would not change, and would do anything to oppose change. And they paid no penalty for that behavior. When this happens, the long-term prognosis for the program, as well as the culture of the company, is not good, and it was not. But for a brief period, a very dysfunctional company did something very special and used the key initiative to nearly achieve greatness. But this clearly shows that for a key initiative to really stick, you have to touch all the bases, especially the last one on institutionalization. You may think you have won, but the Terrorists and the Rebels will fight right down to the end and snatch it away from you. You cannot give them even an inch.

Again, this is an extreme example of what it takes to make a key initiative actually happen, but the takeaways are true for any key initiative and the vigilant reader should pay particular attention to how they are laid out in the next section.

Things to Remember

The six conditions presented in the previous section are a critical requirement for any successful key initiative, and taking a shortcut on any of them will put a key initiative at severe risk. The last case history very clearly illustrates that. However, there are some additional takeaways that the reader should remember. First, key initiatives are a really big deal. As we have said, they require a great amount of commitment in all the resources, including people, time, money, and the collective energy of the company. They are very disruptive. Additionally, the bad people in companies know they are a big deal as well, and they always rise to the occasion to put their best efforts into dismantling the key initiative. So when you start a key initiative, be ready for a war and be committed to win. If you do not want to go through that or you do not want the company to go through that, then do not waste the company's time. Too many top executives make the mistake of underestimating what is needed and then lose heart once they are in the middle of it. The bad people love executives who do that.

Next, it is vital that initially the coalition and later the company have a very clear view of what constitutes the vision. In other words, where does all of this effort lead? In the previous case history, the vision was "zero safety incidents." The Safety VP communicated that right up front and never strayed from it. If you loved the program or hated it, you knew what it was about and where it meant to go. The battlefield was very clearly defined. It is hard to get behind something that is abstract and it is equally hard to make a case for something that cannot be defined in a sentence or two. The vision is, in essence, *a cause*, not just something that is wanted but something that absolutely must be attained. Anything less is unacceptable. The energy required to drive a key initiative can only come from and be sustained by a concept that is not just an idea but is much more powerful—a cause. By the same token, if you expect people to change—not just what they do but what they believe and what they value—then the reason for that change must be up to the task. Those visions need to be crystal clear and very easy to embrace.

This is particularly true for the personal contact required to drive a key initiative. This is not a place for emails, texts, and written manifestos. Those tools have their place, but most of the heavy lifting has to be done the old-fashioned way: face-to-face. People driving key initiatives have to have the energy, conviction, patience, and courage to take their message personally to anyone in the company. Especially those people with influence, who can either be key disciples for the program or very powerful opponents. The SVP and VP in the previous case history told of meetings they attended where the majority of people in those meetings did not just disagree with them but absolutely hated them and what they stood for. They did not shrink from those situations but saw them as a great opportunity to get out the message, understand the issues, know the opposition better, and possibly identify some needed changes in the program. You cannot interact with people effectively if you do not know what they are thinking or what they see as the key issues in the change process. There is a lot of talking in all of this but just as much listening. In the end, many key people saw the commitment of the coalition members, understood the benefits of the initiative, felt that they were listened to and that their issues were addressed, and got behind the program. That only happens with a lot of what people like to call "face time." If you want people to change, respect them enough to put the personal time into helping them do it.

If there is one common theme in this book, it is that change for the better involves a lot of very hard work. You have to be smart about what you do, but then you have to be ready to do a lot of hard work. Key initiatives are in the same vein. You need to have a really good reason for doing one and you have to do it right by staying true to the six requirements. But then you have to be ready to work like hell to make it happen.

Chapter 10

Safety - The First Step up the Ladder to Success

"The safety of the people shall be the highest law."

—Marcus Tullius Cicero (106–43 BC)

Jim: In chapter 9 we said that the actual transformation of a dysfunctional company into a good company involved instituting the right key initiatives as the building blocks that enable a company to take a great leap forward in the journey. In this chapter, we focus on instilling a culture of positive safety and why we believe that safety[6] should be one of the first leadership actions taken to turn around a dysfunctional organization. This is a bold statement, but it isn't theory or an academic observation. As practitioners with decades of experience leading people safely in real world, high-consequence, complex, and unforgiving environments, we know that safety is as close to a proverbial silver bullet as you will find as the lynchpin to transform an ailing company.

---

[6] We use the term "initiative" here for purposes of discussion. In top-notch companies, safety isn't a program and it isn't an initiative. Safety is in the DNA of the organization—a core value defining how things are done as a matter of course.

Safety: The Ultimate Benchmark

What is safety? We like the International Civil Aviation Organization's (ICAO) definition of safety: "Safety is the state in which the risk of harm to persons or property is reduced to, and maintained at or below, an acceptable level through a continuing process of hazard identification and risk management." For our discussion, when we refer to *safety* we are talking about the well-being of people. This is in contrast to pure-process safety (a framework for managing the integrity of operating systems and processes usually in the context of hazardous substances or energy generation). While both facets of safety are essential to good organizational order, our experience is that in a cultural transformation, you must first get people to buy into the change— and you do that by demonstrating to people that they are important to you and that you care about them. Theodore Roosevelt was alleged to have said, "Nobody cares how much you know until they know how much you care." That is certainly the case here.

We have found one absolute behavioral truth—the leadership skills required to successfully manage safety are the same skills needed to manage first-rate operations: attention to detail, focused execution, standardized and disciplined processes, understanding of roles, meaningful metrics, personal accountability, and alignment around the group vision and mission. Nail safety leadership and you will be a long way down the road to a winning culture. Put another way, if you can't lead people in safety, you can't lead people.

A culture of positive safety is a driver of employee engagement—a key constituent of top-tier performance. When people feel that management cares about them as individuals, they willingly reciprocate with differential effort that propels companies ahead of competitors. This can be a major advantage for any business, especially those that are service-driven. To illustrate one example: In the late 1990s, there was an effort by a CEO to establish a more just culture in a major, highly unionized transportation company. As with most companies in this industry, the cultural inertia was adversarial. Conflict and skepticism had evolved on both sides over decades. It was a poisonous atmosphere with pervasive mistrust between company management and their employees and union leaders. The toxicity was palpable, and cynicism was manifest on each side. Needless to say, attempting to transform a culture in such a hostile atmosphere was a daunting task. There was only one real common-ground issue that could be used as the initial "flagpole" upon which everyone could rally: *safety*. With safety as the baseline value, the effort moved forward, starting with some candid conversations around the central issue that was causing many of the battles: the company discipline policy around safety incidents. In a nutshell, as constructed, the long-standing discipline process was viewed by unions and employees as unfair, unjust, and overly castigatory for workers, especially for simple mistakes made while trying to do the job. This *one-size-fits-all* policy imposed punishment, with little regard for whether the offender was a willful, neglectful employee or whether the offender was a good worker just trying to do the right thing who, in the process, erred. The policy's effect was the alienation of employees, and the workplace was awash in mistrust. People, in turn, felt forced to hide safety issues for fear of discipline. The company was not a safer place with this policy because safety issues were driven underground. All in all, it was a loser for all stakeholders.

The order of business in the cultural makeover was to abolish the discipline policy and start over. Rather than the traditional approach where human resources managers and/or lawyers crafted the discipline policy, this innovative CEO commissioned Jim to recruit and lead a multidiscipline team. What was most remarkable about the exercise was inclusion of union leaders as full partners in development of the new policy. What's more, the CEO ordered all existing discipline records cleared, in effect giving everyone a fresh start under the new doctrine. Not surprisingly, there was much wailing by many in the management ranks, with claims that the sky would fall and the company would implode. Neither happened. In fact, this initial gesture left little doubt that the CEO was serious about changing the paradigm.

The results were incredible. Within three years, the culture dramatically turned. Labor unions and employees were far more aligned with management's vision and commitment to customer excellence. Safety improved, service improved, costs went down, and people liked coming to work. Industry press picked up the story. The rest of the industry was watching closely.

This simple story makes the point: One day deep into the change, an employee crew was scheduled to deliver a load of lumber to a small-volume customer. When the crew arrived at the plant, the entry gate was locked with nobody in sight. In the former adversarial atmosphere, the employees would have shrugged and simply returned with the load to the terminal with the attitude that "It isn't our problem." However, with the new, more collegial collaborative culture, the employees took immediate ownership of the problem. They took time to find a telephone number of a plant supervisor to open the gate. Plant workers were so delighted with the employees' actions that the site manager called the company to offer praise. It seems that the shipment arrived just in time or the plant would have had to shut down and send people home. On the surface, this may seem like a small, anecdotal story. But it is much more than that—it is evidence of how deep the change migrated into the working face of this very large, diverse, com-plex one-hundred-plus-year-old company. This kind of differen-tial effort started to happen across the enterprise, as people felt included and appreciated, and they felt ownership. Remember, this transformation all started with leadership's focus on safety as the cultural North Star to heal internal strife.

As illustrated in the preceding discussion, with commitment to safety and just treatment, management sends a very strong message to employees, customers, and communities: we care for the well-being of our people. When your people believe that you care, they will willingly extend constructive differential effort that brings both tangible and intangible benefits to a company. They invest personally in the organizational mission and become posi-tive brand ambassadors. And that makes for a powerful competi-tive force that is impossible for competitors to replicate. Costs are reduced, safety improves, equipment lasts longer, and customers are treated with additional respect. This garners brand loyalty, which often translates to customer willingness to absorb reasonable pricing increases. It also provides more willingness to forgive mistakes or occasional gaps in service. Just as important, engaged customers will recommend your company to their friends and others in business.

Customers: Why Safety Matters

Ignoring safety will cost your business long-term in the modern era, and putting safety on the back burner will hinder your ability to grow. Smart customers now perform due diligence on their vendors. In our experience, more and more good companies, the kind you want as customers, are basing their business decisions on potential partner safety performance. Reputation, brand quality, and service are big deals to customers. Few customers want their brand linked with poor-quality performers. Safe, clean, responsible conduct—not necessarily price—will often be the element that determines if a customer will hire you or someone else. They select vendors/suppliers carefully and will pay a premium to do business with safe, cordial, and efficient companies. They also want a vendor partner who is prepared if anything goes wrong: proper insurance levels, transparent communications, and detailed root cause analyses. A healthy safety culture is job security at this most basic level for those who manage customer products and reputation. Put yourself in the position of a customer: What really matters if your business survival depends upon getting your product to your customers safely and on time? Is it a bargain shipping rate, or is it value-added service that is safe, efficient, and responsible?

It is difficult to provide a safe, cordial, and efficient customer experience if your frontline people are not engaged and aligned around your safety vision. Your employees understand that an investment in safety is an investment in their well-being. Putting employee safety front and center demonstrates management's commitment to people. Show us a world-class, highly productive, and profitable company, and we'll show you a company that cares for employees with safety as the cornerstone value.

Culture and Safety

As we pointed out in chapter 8, culture is the bone marrow of an organization. It is the adhesive that joins an organization together. Research and experience have established that culture can be the ultimate differentiator in any enterprise. As we stated earlier, an excellent culture is one where people and process are in harmony with the company's vision and values. Accepted patterns of behavior are clearly expressed and understood and, ultimately, customers enjoy positive and consistent experiences in their interactions with the company.

An energetic culture will not develop in an autocratic environment where decisions flow from the top down. Employees who are not empowered are simply not engaged in their jobs. At its core, a company's culture is behavior-based. It is a mix of individual and organizational behavior, all of which is functioning in alignment with a company's values. In a traditional, top-down command-and-control environment, employees lack ownership of their work and do not have a connection to the business. They have no stake in the game. They will disengage.

Some Keys to Great Safety

We have defined several actions that helped us formulate successful transitions in driving a culture of positive safety. While this listing is not all-inclusive, it represents major leadership efforts that, if observed, will accelerate the journey to a healthy safety culture. The actions are:

The CEO must drive safety—In chapter 6 there was a lot of discussion around the role of the CEO in leading the transformation. We said that the CEO has to take the lead and be responsible for all the parts of the transformation. That applies in spades when it comes to safety. To reiterate a key point from chapter 6, unless the CEO can inspire ownership and involvement in the hearts and minds of management and the frontline workers, positive safety change will not happen. Of course, as we said in chapter 6, the CEO will have people actually running the day-to-day, but there can be no equivocation—the CEO must be seen as the ultimate chief safety officer through his actions, words, and visibility.

Make safety a core value—We hear organizations all the time say, "Safety is our top priority." Well, being a *priority* just isn't good enough. Safety must be a core value of the organization. Knowing the difference between values and priorities is a major point of confusion for many organizations. Values define and bind the team together, drive decisions in every interaction, and never change. On the other hand, priorities are tasks requiring action, must be managed daily, and shift frequently based on a given situation. We must ensure that values are never compromised. Safety and safe behaviors must be nonnegotiable values.

Pay attention to detail and do the basics well—That means you need to have disciplined processes, standardized procedures (SOPs), and laser focus on proper execution. These qualities structure the culture and provide the tangible order necessary to drive results.

Ensure that your culture propels safety—In a healthy company, the culture powers the safety process—not the regulators, not the lawyers, and not the media. Contrast this to mediocre companies where management focus is on compliance alone. Here, the lawyers generally prevail and drive the program with the goal of not doing anything more than is necessary for fear of perceived legal vulnerability. This, at best, produces unexceptional minimum safety results. Regulations do not always consider the value of a proactive, forward-leaning safety leadership approach. For example, a regulation may require that truck drivers, locomotive engineers, airline pilots, or nuclear plant operators have a minimum number of hours off-duty. The intent, of course, is to provide time for rest. The assumption is that the human will use time away from the job to rest and report back at the end of the statutory period rested and ready to go. Few regulations explain why it is important, nor do they promote understanding of circadian cycles, sleep physiology, work-rest dynamics, signs of fatigue, or countermeasures to help induce sleep when off-duty.

Remember that you are on display—Managers in organizations with positive safety cultures understand that employees watch for cues on what is important. People will do what they think is important to the boss, not necessarily what is said or what is written in policy statements. "The safety behaviors and attitudes of individuals are influenced by their perceptions and expectations about safety in their work environment, and they pattern their safety behaviors to meet demonstrated priorities of organizational leaders, regardless of stated policies," said Dr. Dov Zohar as cited by NTSB's Hon. Robert Sumwalt in his presentation at the Patriot Rail Company senior leadership team workshop in November 2013.

Measure performance—As in all critical functions, safety needs to be measured. There are a few cautions we can offer here. First, in bad companies, there is undue emphasis on keeping score ....focusing on backward-looking outcomes of safety not forward-looking prevention. Such an approach is, in effect, looking at yesterday instead of looking forward to today or tomorrow. Bad companies watch only the numbers and pay little attention to what drives the numbers. It is not an effective way to win. It is akin to playing a baseball game with everyone on the field staring at the scoreboard—and paying no attention to hitting the ball and running the bases. Good-performing companies also keep score, but they do it as part of a total approach. There is clear recognition that it is the actions and processes that drive the outcomes. So they measure other things that drive safety, such as management involvement in safety events, safety meetings, observations, and monitoring findings, to name a few.

Keep at it—Great safety is a journey, not a destination. Be obsessed with improvement. Don't let up. Remember that the journey itself forces improvement. And don't let occasional setbacks or naysayers get in the way. It is sometimes hard to keep the bean-counters and other skeptics engaged since, as Dr. James

Reason puts it, "Safety is a dynamic non-event; we have to work very hard so nothing will happen." When nothing happens, some people lose interest in staying hard at the task. Dr. Tony Kern said it well in his book *Darker Shades of Blue: The Rogue Pilot*: "A safety culture must be inspired and constantly nurtured to prevent that downward spiral into disaster."

Listen to the organization—We have learned over the years that an organization will talk to you. It may be a series of close calls in safety, or it may be an accident. It could be high turnover in the ranks or growing discipline problems. It might even be your gut as a seasoned leader telling you that things are not right. All these can be indicators that dysfunction and discontent may be sneaking into your organizational blood flow. Whatever it is, you need to be tuned in, listen, and act. One way to ensure you hear what is going on is for management to regularly conduct "start-stop-keep" discussions with the front line. In this exercise, ask your people: "What aren't we doing that we need to start?" "What are we doing that we need to stop?" and "What are we doing that we need to keep doing?" You would be surprised how effective this exercise is in helping you keep your finger on the group pulse. It also enables the front line to feel included and demonstrates that they have impact. You benefit by getting some new ideas on how to better accomplish the mission while eliminating non-value-add activities.

Instill a culture of openness and justice—This is an especially critical element in creation of a positive safety culture. One tool some companies are adopting to demonstrate commitment is issuance of a non-reprisal policy that formalizes the strategy and clarifies organizational and individual responsibilities. This is one template that we have used:

XXX Company Non-Reprisal Policy (sample)

- Our value. Safety is the cornerstone value at *xxx company*.

- Our goal. Our goal is zero injuries, accidents, and casualty events.

- Our commitment. We are committed to be the safest organization possible. Any task that cannot be done safely should not be attempted until it can be done safely.

- Open reporting. It is imperative that we have open, free, and good-faith reporting of any hazard, occurrence, or other information that in any way could affect the safety of operations.

- Individual responsibility. Every individual at *xxx company* is responsible for acting safely and reporting to any supervisor or manager information that may affect personal or process safety.

- No reprisal. To promote the timely, uninhibited flow of critical safety information, this process must be free of reprisal. Accordingly, *xxx company* will not use this reporting system to initiate administrative or disciplinary action against any individual who discloses, in good faith, information on a hazard or occurrence that results from conduct that is inadvertent, unintentional, or not deliberate, and that is not a pattern of behaviors or repeated misapplication of rules.

- Expectation. We expect that all *xxx company* personnel will endorse this program to help ensure our company continues to provide employees, customers, and communities the highest level of safety.

<u>Institute and follow a Safety Management System</u> (SMS)—Safety practices generally circulate in three main arenas: technical, human, and organizational. One disciplined approach to ensuring all three realms are addressed is to implement a tailored SMS plan. There has been some confusion around SMS—just what it is and isn't. For our purposes, as defined by ICAO, SMS is an organized

approach to managing safety, including the necessary organizational structures, accountabilities, policies, and procedures. In substance, an SMS program establishes processes to collect and analyze data on potential safety problems and then evaluates mitigations to resolve the safety risk before an accident happens. According to ICAO, the major components to an SMS program include four pillars:

- Safety Policy—Establishes senior management's commitment to continually improve safety; defines the methods, processes, and organizational structure needed to meet safety goals.

- Safety Risk Management—Determines the need for, and adequacy of, new or revised risk controls based on the assessment of acceptable risk.

- Safety Assurance—Evaluates the continued effectiveness of implemented risk-control strategies and supports the identification of new hazards.

- Safety Promotion—Includes training, communication, and other actions to create a positive safety culture within all levels of the workforce.

Accountability—Throughout this book, we have discussed widely the concept of accountability. When it comes to safety, a lack of accountability is high on the list of safety culture killers. In our experience, there is simply no way to build a successful organization without solid accountability throughout the ranks. We believe that accountability is manifest most prominently in three realms:

- Individual accountability—Here, the individual accepts accountability for his actions. Individual accountability is most richly cultivated in a just culture as discussed in chapter 8, where people are treated with respect and inadvertent errors are not punished (as opposed to willful, deliberate risk-takers who must be dealt with firmly).

- Organizational accountability—Here the management lets go of a traditional blame culture and accepts that the organization possibly has a role in creating an environment that contributed or led to the failure. Is there sufficient training? Did we clearly outline the right SOPs? Did we provide the right tools? Did support systems function properly? Did we address employee concerns? Have we taken the right management actions to ensure clarity around our expectations? Did we create an atmosphere where normalization of deviance was accepted? Did we do everything to ensure that the employee is successful?

- Peer-to-peer accountability—Here, the individual recognizes and accepts his or her responsibility to work safely. In a healthy safety culture, individuals feel free to step in and confront poor safety behaviors. This is only possible when there is an absence of a blame culture cultivated by management. In one company we know, the program was called "tap on the shoulder." In this program, employees felt free to tap a fellow worker on the shoulder if they saw something amiss. Everyone knew that it was an expectation. When people police each other in the workplace, you are well on your way to a productive safety journey.

Some Common Noxious Behaviors That Kill a Culture of Positive Safety

Good leaders understand that there are highly damaging organizational behaviors that, if allowed to root, will derail the journey toward a healthy safety culture. Some of the common safety culture killers that we have seen include:

Ignoring the facts—One thing that distinguishes good companies is the ability to have a balanced view of themselves and their realities. Good leaders do not ignore the facts, even if brutally unpleasant. They recognize the need to be obsessed with continuous safety improvement. They also recognize that complacency creep is real. Safety can be an unforgiving and uncompromising domain that grants no waivers. Safety success in the past doesn't guarantee safety success in the future. There is no credit for how good you were in the past. Safety doesn't care what you did last year, last month, last week, yesterday, or even one minute ago. Safety cares only about right now. It demands full attention every second. And safety is unfair. Think about it—if you live to eighty years, you will experience more than 2.5 billion seconds of life. But a mere handful of seconds of distraction or poor decision-making is all it takes to dramatically alter or destroy the rest. Now that is unfair. What makes it worse is that there is no choice as to time of reckoning when a failure may occur. Catastrophic failure may sit dormant on the shelf for a long time and then strike when least expected. Safety risks may not always be where you think they are.

Delegating safety to the safety department—Safety is a management responsibility, not the responsibility of the safety department. Safety people are coaches and mentors. You can't delegate safety—you own it as a leader in your company. This fact was clearly illustrated in a company we know that had more than three hundred safety managers on the payroll, yet the company posted an appalling safety performance. An assessment of the company affirmed that the company operating and executive management had zero accountability for safety performance of their people. In effect, they gave themselves a pass by abdicating responsibility to manage and lead their people safely to the safety manager. If there was a safety problem, they just added another safety manager. The safety managers were also loaded with administrative duties that had nothing to do with coaching, mentoring, and training people in the safe performance of their duties. And these safety managers were not respected or empowered.

Not recognizing that "good" can be "bad"—Our thesis is that "good" can be "bad." By that we mean that good performers, be they individuals, groups, or companies, must guard against a false sense of invulnerability and complacency. Put another way: good performance, and an absence of accidents, can promote complacency—satisfaction with one's accomplishments, accompanied by a lack of awareness of actual dangers or deficiencies. Wilber Wright, in 1901, said it very effectively: "Carelessness and over-confidence are more dangerous than deliberately accepted risk." To illustrate, there are hundreds of transportation accident reports that prove that even the most experienced, knowledgeable, and capable professionals are not immune. Unfortunately, there are some fairly recent dramatic examples, such as Asiana Airlines Flight #214 that crashed on approach to Runway 28L at San Francisco International Airport in July 2013. As reported in the press and affirmed in the National Transportation Safety Board (NTSB) investigation[7], the airplane was crewed by three highly experienced, highly trained, and seemingly highly motivated pilots, flying the latest state-of-the art Boeing 777 airplane. As documented in the NTSB report, despite thousands of hours of combined experience as pilots, the airplane dropped too low and too slow on approach. The main gear wheels and undercarriage struck the seawall prior to arrival at the runway threshold, causing severe damage and loss of control. The airframe was destroyed, 187 people were injured, and 3 people died. This is just one illustrative example. There are countless more we could quote to attest to the fact that people of high skill and practice have found themselves involved in a breach of safety in many industries.

Declaring victory before getting to the finish line—You may think you are good, and you may be ahead of other groups or your competitors in your safety, productivity, or financial numbers. But beware of premature declarations that you have won. You haven't. Dr. James Reason made the point very clearly: "There has to be a sense of chronic unease when it comes to safety performance." His point—you should never think that you are *okay* in safety.

---

[7] NTSB Aircraft Accident Report (AAR-14-01), "Descent Below Visual Glide Path and Impact with Seawall: Asiana Airlines Flight 214; Boeing 777-200ER, HL7742."

Tolerating toxic people—As we discussed in chapters 6 and 7, toxic people (the Terrorists and Rebels) can sabotage good social order and team alignment. These people rob positive emotional and intellectual energy from the good people. T. S. Eliot said, "Half the harm that is done in this world is due to people who want to feel important. ....they are absorbed in the endless struggle to think well of themselves."

IWHTM syndrome (It Won't Happen to Me)—Some people believe that they are bulletproof and immune to calamity. If you have someone with the IWHTM syndrome, you have someone who is reckless and may be out of control. This attitude leads to trouble, especially in high-consequence environments. Safety databases are full of accident investigations where someone caused bad things to happen because they believed IWHTM.

Losing respect for SOPs (standard operating procedures)—In high-consequence activities, people must follow standard procedures, rules, and practices. Freelancing and winging it are risky. One of the most common examples of this phenomenon may be the notion of selective compliance. Here, people choose which rules and procedures to obey and which ones to ignore. The hard truth is that if you want a culture of total safety, there must be respect for all rules. If a rule doesn't make sense, eliminate it. So why don't people follow SOPs? We think there are five main reasons:

- The organization may not have adequate SOPs.

- There is poor organizational commitment to SOPs—individuals let up.

- There is intentional disregard of SOPs.

- The SOPs lack credibility and therefore are ignored.

- The organization has not effectively communicated SOPs in a way in which they are understood at all levels.

<u>Falling victim to the normalization of deviance</u>—This is when people within an organization become accustomed to shortcuts in the conduct of their work. Over time, with no consequence for the deviations from accepted practices, the shortcuts become the norm. There has been much written on this topic, but the theory gained prominence in good measure due to Diane Vaughn's 1996 book *The Challenger Launch Decision,* in which she details, in her view, how the normalization of deviance ethic at NASA resulted in a design flaw in certain performance capabilities involving joint seals on the solid rocket booster. This defect resulted in failure shortly after launch and subsequent loss of Space Shuttle Challenger and crew on January 28, 1986. Vaughn pointed out how "evidence initially interpreted as a deviation from expected performance was reinterpreted as within the bounds of acceptable risk."

Mike Mullane, veteran NASA astronaut, aviator, author, and motivational speaker, is a highly respected expert in normalization of deviance. With decades of experience in space flight and military aviation, Mullane clearly makes the point, "Budget and production pressures often force people to take a shortcut. Here's the problem: the shortcut will usually work. Because it worked, team members will be tempted to take the shortcut repeatedly, and the shortcut becomes the norm. As a common example, an employee can get away without wearing personal protective equipment, and nothing will generally happen—until that one tragic day."

One of the most obvious current-day examples of normalization of deviance that we can think of is the use of personal electronic devices (PEDs) while operating a vehicle. Virtually everyone we know admits that using PEDs/cell phones while driving or operating moving equipment is risky. Yet even with laws in many jurisdictions prohibiting such use behind the wheel, people still do it. It has become the norm.

<u>Overcomplicating things</u>—"When you hear hoofbeats, don't start looking for a zebra. It is probably just a horse." We like that well-known aphorism because it points out the folly of confusing even

simple concepts. Some people have a penchant to obfuscate everything. They put too much emphasis on process and little on outcomes and content. There is a flurry of activity but little substance comes out the tailpipe. In our experience, safety is all about executing the fundamentals. If you do the basics well, everything else will normally follow. No need to try a triple-reverse pirouette off the high dive to get wet. Just jump into the water.

Getting lost in the numbers—This concept was mentioned earlier and is directly applicable here. Measures should encompass process and execution, not just outcomes. If you successfully execute the basic processes, the right results will follow. Also, we know that metrics must be clear, direct, and easy for everyone in the organization to understand. Patience is required because when you undertake safety culture renovation, the safety numbers may worsen before they improve. The increase in reporting likely means that you are getting better information and a more accurate view of the real underlying risks as people begin to trust the process and employees report things that before were hidden and left unaddressed.

Trying to perfume the pig—A pig is a pig, even if you try to disguise it with perfume. So it is in some companies. Managers try to disguise problems through what we call *happy talk*. These are people who extend effort to avoid candid conversations. In the process, they redefine *good* to avoid the truth. Constructive conflict is an important and essential part of organizational life. Trying to put perfume on a problem won't make it go away—it just temporarily masks it. It is best to deal with it head-on.

Forgetting who has the power in the organization—Who would you say is the most powerful person in your company? It is likely that you may immediately respond "The CEO" or "The president." In our view, however, the most powerful people are the ones who have day-to-day domain over resources and are the point of direct contact with the customer. To be sure, the CEO, president, and senior managers have authority. But do they really have the power? Recognizing the difference between power and authority is an important attribute of great leaders and great companies. Power resides in the front line. Few CEOs, presidents, and senior managers can do the tasks necessary to meet customer expectations. In addition, your front line knows where real risks lie. They are the subject-matter experts. They have the answers.

The Safety Crucible: Leading People Safely Home— Everyone, Everyday

Is there a nobler professional calling than to lead others? Is there any higher responsibility than to ensure we lead people home to their families? Greek philosopher Heraclitus (c. 535–c. 475 BC) is alleged to have said, "Out of every 100 men on the battlefield, 10 should not be there, 80 are nothing but targets, 9 are fighters and we are lucky to have them for they the battle make. Ah, but the ONE. One is a warrior, and he will bring the others home." Leadership requires a warrior ethos, and, as leaders, bringing people safely home is our duty, our right, and our obligation.

Creating a positive safety culture is a leadership responsibility and not something that can be delegated. If you, as a leader, make safety a low priority, the people around you will follow your lead. If safety is confined to a department, it will never take hold throughout the organization. A safety department can coach, train, and mentor. Only an operationally aligned organization can execute on safety with excellence. Safety execution requires a different and vastly broader approach than regulatory compliance. It must be driven by company leadership and then anchored to the front line.

In a 2007 address to the Air Line Pilots Association International Safety Forum, the Honorable Robert L. Sumwalt, NTSB, said, "[As] leaders, I suggest that you not only have the ability to influence safety, but you have the obligation to do so, as well. Speaking up and saying something about deviations is hard. It's unpopular. But, remember, all progress has resulted from people who took unpopular positions. If you accept anything less than standard, you send a message to others that it is okay to perform to a lower standard."

As leaders, we have great capacity to directly affect lives. Not only do our decisions and actions involve our employees, but they also affect others close to those employees—spouses, children, parents, siblings, friends, and neighbors. It is, in the most basic terms, a privilege to lead people. However, it is also a formidable responsibility to create an energetic and positive culture of safety, especially in high-consequence scenarios.

Remember too that there is no finish line. You are tested repeatedly on your ability to lead others. There are two main levers with which leaders must effect change and drive performance: people and process. We have fifteen management truths that serve as foundational elements in building and sustaining a culture of positive safety.

1. <u>This is not a grassroots affair</u>—A culture of positive safety must be leadership-driven but employee-based. Waiting for a healthy safety culture to bubble up from the ranks is an ineffective strategy.

2. <u>This won't happen by chance</u>—Just like any other major program, instilling and sustaining a culture of positive safety requires a deliberate plan.

3. <u>A great safety culture is a journey</u>—It is not a destination. You will never achieve total safety. The benefit to the organization is achieved in the journey itself.

4. <u>Safety must be a core value of the enterprise</u>—Priorities change. Values don't change.

5. <u>Management must have courage to stay the course</u>—Disappointments will happen in the journey. Naysayers will be quick to claim defeat. Don't let it happen.

6. <u>Zero is the only acceptable goal</u>—Although 99.9 percent is a pretty good performance standard in most business arenas, it's not when it comes to safety. As outlined in

chapters 8 and 9, *zero* is a mind-set and the value comes from the journey to zero and the quality processes it forces on managing safety and gaining buy-in at all levels.

7. <u>There must be organizational accountability for safety failures</u>—Blaming individuals at the front line without examination of organizational issues is the hallmark of unenlightened management.

8. <u>Safety goes beyond compliance</u>—Regulations, rules, and laws are the baseline. Great safety requires individual commitment and personal accountability.

9. <u>Leaders should focus on execution, pay attention to detail, and promote high standards of behavior</u>—Confucius said, "Life is really simple, but we insist on making it complicated." Execute the basics well and keep your behavioral standards high, and you have the battle nearly won.

10. <u>The focus is on at-risk behaviors not per-se conditions</u>—Addressing behavior, not focusing exclusively on deficiencies in equipment or conditions, can prevent many casualties.

11. <u>Safety is the ultimate organizational North Star</u>—A foundational safety culture can help moderate typical internal and external distractions. It is the common ground around which people may freely rally to align with the mission.

12. <u>Safety success in the past doesn't guarantee safety success in the future</u>—There must be a leadership obsession with continuous improvement or the organization risks stagnation.

13. <u>Safety requires solid SOPs</u>—Ensure SOPs are meaningful, clear, and followed.

14. <u>A low-cost solution</u>—Transforming safety is a low-cost solution. Big capital is not needed. It is about thinking and acting in a new way. Costs that are incurred to fix broken things are an investment in the culture. We must reject the old-school thinking that safety is a cost or somehow hinders production. If you think an investment in safety is too expensive, consider the opposite. Ignoring safety can ultimately lead to accidents, litigation, increased insurance cost, product failures, damaged reputations, poor employee morale, and eventually business failure.

15. <u>It is not safety vs. productivity</u>—We know that safety drives high productivity. To argue otherwise is folly. In most operations, downtime from a safety failure is expensive—both in direct and indirect expenditures, and is much costlier than is the leadership attention and investment it takes to prevent the disruption. Safe operations actually make a business more profitable with less rework, higher equipment availability, added flexibility to deploy resources where needed, and, perhaps most of all, trained, confident employees who stay on task with disciplined execution. People retain focus on the right behaviors so there are fewer do-overs and mistakes that destroy production.

We know that building a better business culture relies on engaging and empowering employees. Their actions and behavior drive the culture. Leaders set the course for a healthy culture, but employees bring it home. It's the same with building a culture of prevention. No matter what we say or do to align our company with the understanding that safety is a core nonnegotiable value, it doesn't matter unless our front lines are fully with us.

All of this begins and ends with leadership. The team must sense your urgency. They must see you focused and concerned every time you have a safety situation, severe or not. You must perform a root cause analysis on every issue, not just those with a costly outcome. You must have a continuous obsession in your organization to prevent complacent behaviors from setting in. In short, safety success depends on leadership ability to stand above the crowd and lead with conviction. Here are fifteen daily leadership principles that we have used to effect change and sustain a culture of positive safety.

Fifteen daily safety leadership principles

> *"Leadership is about influence. Nothing more, nothing less."* —John Maxwell

1. **LEAD**. Don't delegate safety. Be an example. You are on display every day.

2. **EXECUTE**. Focus on execution of fundamentals.

3. **MANAGE**. Manage your high-risk situations and top three causes of casualties.

4. **COMMUNICATE**. Communicate your commitment and back it with actions.

5. **SEEK INPUT**. Encourage employee input. Take time to listen.

6. **FOLLOW UP**. Follow through on commitments and keep people posted.

7. **BE RESPONSIBLE**. Accept responsibility for the safety performance of your team.

8. **MAKE SAFETY PART OF YOUR DAILY PLAN**. Make safety an essential part of your business plan.

9. **COACH**. Mentor and lead employees in safety.

10. **REINFORCE**. Positive reinforcement gets everyone involved.

11 **PROMOTE ADVOCACY**. Seek out and reward safety advocacy.

12. **BE CONSISTENT**. Safety is a daily focus—not just when convenient.

13. **INSTILL A JUST CULTURE.** Advocate honest reporting of errors without reprisal.

14. **SEEK ROOT CAUSE**. Support root cause investigation findings, even if painful.

15. **PUT PEOPLE FIRST**. Be hard on facts and easy on people to build a culture of trust.

Closing Thoughts

At the beginning of this chapter, we made the point that safety is a cornerstone element in any cultural transformation. We said that people need to know that leaders care about them before they will buy in to or align with a leadership vision—and that nothing better demonstrates that leaders care about people more than undertaking a journey toward a culture of positive safety.

Once the safety journey is well under way, you are ready to tackle two other primary initiatives: building an effective HR function and instituting a robust performance-management system. These three—safety, HR, and performance management—are the three legs of the stool that must be standing to most efficiently drive constructive change in any company. In chapters 11 and 12, we examine both HR and performance management in detail.

Chapter 11

The HR Function

Misunderstood and Underutilized

Chuck: In the previous chapter, we explored one of the three pillars of a functional company: safety. This chapter on HR and the next chapter on performance management complete the three key initiatives that are absolutely necessary to get a company moving in the right direction. Once a transforming company provides a safe environment for its workforce, hires and prepares the right people, and then provides them with effective management so that they can excel, then they are in a position and have the necessary capabilities to move forward on what they deem are the next areas to be addressed. But that is for later. What we are focusing on here is putting the priority on people and building the necessary infrastructure to attract the right people, get them on the team, and retain them.

As was pointed out in chapter 2, bad companies place little value on their people, which is amazing given the huge investment in human capital that virtually all companies must make. They fail to realize that companies need high-quality employees who are well trained and well managed if they are going to excel. Bad companies just do not put much effort into building the team. The cause can be ignorance, arrogance, laziness, or any combination of the three. The feeling seems to be that it is only necessary to fill open positions with somebody and it will all work out. Of course, nothing could be further from the truth. Good companies know that building the team and maintaining the team are top priorities. Their managers are well versed at understanding their needs, spotting the right talent, and attracting that talent to the company. Once new people are on board, they are given the appropriate training and mentoring to prepare them for their positions. The goal-setting, reviews, and other performance-management tasks that follow are first-class and offer the new employees a real opportunity to excel and build successful careers within the company. Good companies tap all of the potential that their people have to offer. And they have a very strong and capable HR (human resources) department to manage and lead the effort. That is why they are good. Bad companies seem to have little interest in human potential, usually cannot identify it, and generally drive it away when it lands in the company by accident. To them, HR is a do-nothing department that should be ignored. That is why they are bad. Jack Welch once said in an interview that a good strategy could fail if executed by a bad team but that a bad strategy could be successful if implemented by a good team because that team would make the necessary adjustments to make it work. It is all about having good people.

Most CEOs brought in to turn around a dysfunctional company understand something about the basic people problems in bad companies. However, in many cases, even these CEOs do not completely fathom the total lack of appreciation a bad company has for the importance of human capital. These CEOs often end up starting in a much deeper hole than they assumed, and this big surprise can sometimes be enough to derail the whole transformation effort or at least keep it from being more than just partially effective. It is vital that these CEOs get the people part of the transformation right. To start that moving in the proper direction, they need to do a really good assessment of what they have, how it got there, and how it is managed before they can begin to plan and make changes. The next section presents what a new CEO can expect to find in the way of people, people processes, and what masquerades as an HR capability. The rest of the chapter goes into how to make the necessary improvements, including those in HR, to build some functionality in the people area that will transform not only attitudes but performance levels. This is not only a key initiative but, as was previously stated, one of the three most important key initiatives. You cannot get substantially better if the people part of the equation is absent. The critical thing to remember here is that most major transformations fail at the people level. CEOs do not fully appreciate what they are up against on the people issue and underestimate the needed effort, make basic mistakes in evaluating the team they are starting with or in creating and implementing the required program changes, or are too weak or too slow in dealing with the issues and/or needed corrections. Whatever the reasons, CEOs have to get this piece right or the whole effort will be derailed.

What You Typically Encounter

In chapter 2 we went over the culture of a bad company and what that company values. Managers in bad companies do not value people or the idea that great people can do great things, but what they do value over everything else is control of their environment and independence. Resistance to change and fear of processes are not far behind. This culture that ignores people, values independence, and opposes any form of change creates the mediocrity that you persistently encounter in the dysfunctional company. When there is little interest in attracting or developing quality people, what you end up with in these bad companies is ineffective management and mainly substandard people, and some unidentified and unappreciated good employees who wonder why they are there. Managers are too busy failing with weak and untrained employees and equally poor management methods to have time or interest in upgrading their team. People are generally seen as expendable and interchangeable rather than as potential game changers. The idea of building a world-class team to do world-class work is lost on bad companies. And since the leadership in bad companies would rather fail doing it their way than consider a different approach, it is going to take a massive campaign to break out of this vicious circle and build something positive.

Given that people are not highly valued by bad companies, it follows that HR is seen not as a core activity but as just a secondary support function with a limited role and not much influence. The HR head is usually considered the weakest member of the senior leadership team, if he is even on that team. It is often a good place to put someone who is not very talented but has politically *earned* an advance to that level. In the HR position, this person has some minimal status but cannot cause too much damage or get in the way. It is usually best practice within the bad company to put someone in that position who is just happy to be there, knows their place, and does not expect much in the way of opportunity. To say that the head of HR is not a respected leader is often an understatement.

The HR head pretty much adopts a hands-off approach when dealing with other departments, especially the field. Most senior managers and their people do not want any help from HR, and the HR head does not have the confidence or the interest to get involved in their affairs. In fact, there is a general reluctance by the HR head to take the initiative on much of anything, especially personnel issues that affect the front line. An HR group member at any level who tries to influence other managers on their personnel activities is probably putting his job at risk. A manager may ask for help on a termination issue, have questions about benefits, or need some assistance on a training idea, but such requests are rare and do not typically require deep involvement from HR. The HR group and the rest of the company generally ignore each other, and everyone is fine with that. An HR SVP in a large service company once remarked, when asked if his department tracked personnel turnover, that he saw no value in it and did not think it was his responsibility. He felt no ownership or accountability for the quality of the company workforce or how well it was managed. He legitimately thought that none of that was part of his job.

Incumbent CEOs in bad companies create the environment that results in this limited and ineffective HR function. These so-called leaders are firmly rooted in the bad company culture. In their world, HR has always been a second-class organization with few responsibilities. With no CEO focus on, or interest in, building a company skill set regarding any part of the people process, there is no message from the CEO that the HR department is anything more than weak support. Standards for hiring, workforce gap analyses, training needs and effectiveness, and goal-setting and reviews are not part of the charge to HR from the CEO. If they were, the HR group would not be up to it anyway. If this is going to change, it takes not only a new CEO but also replacement of all the senior leaders who are not open to a new culture and a new approach to building and managing the employee base.

In dysfunctional companies, personnel issues are generally handled locally on an *a la carte* basis. The local managers prefer it that way, even though they have little or no interest in anything that is related to employees. They want the control. There may actually be people who focus on personnel issues trying to do the work for local managers, but they usually are not trained, are not particularly talented or supported, and are working in a vacuum doing what they think is best. It may be possible to find some fairly good HR-type programs in a particular business location, but they are isolated and the local people are doing it all on their own. There is no overall strategy or process for hiring and managing personnel. All managers do what they want or what they think they have to do to address a particular personnel problem. The only directive from the corporate office is to "keep it legal." Pretty much everything is reactive. As long as every position is filled, there is no reason to focus on employees and their needs. Personnel issues are seen as a nuisance that should be handled quickly so that everyone can go back to doing important work—which is usually whatever they feel like doing. The leadership and management in bad companies generally have little experience with any kind of a quality program for people. They do not understand it, are not good at it, and really want to avoid it. As a result, they have no idea what they need or want when it comes to personnel. They typically do what is easy, which means keeping people regardless of performance, hiring friends or people who walk in the door, or leaving positions unfilled. If there are any positive activities such as goal-setting, performance reviews, ranking of people, or skills training, they are usually done locally by someone with good intentions but little experience. These "one-off" efforts are better than nothing, but they have no broad support, which means they reach few people. Corporate attempts at these activities are rare, usually of indifferent quality, and do not last long. The HR team does not know how to do it, and there is no real support for it anyway. Bad companies are in denial about all of this. They are locked into the "myth of good" and think they have good people who are happy to be with the company. When people who are actually good become frustrated and leave, company management does not know anything about it because they do not believe in exit interviews.

It is rare to find any leaders or managers in a bad company who feel any ownership for the process of hiring and managing employees or see it as particularly important. Managers may do it but, like the HR SVP mentioned previously, they do not feel they have any accountability for the quality of the people they employ or how well they are managed. The people process is basically an orphan. A logical person would ask why the HR group does not just fill the vacuum and take control of this disorganized mess, pull it together and make it consistent, and create some capability within the company for developing and managing talent. The bad company culture will not let that happen. Management actually thinks that the only role for HR is to administer benefits and perhaps help out, when asked, with personnel matters in the corporate office. They see HR as a department that you have to have but do not need to use. Managers in bad companies do not like to give up control of anything, admit they need help, or accept input from anyone considered an outsider. Even though they find personnel matters repugnant, they would rather fail at them than ask HR for help. They just do not see it as an option. For them, HR needs to stay very small and almost invisible. And it follows like safety, as we saw in the previous chapter, that when the HR department also sees itself as small and not as a critical part of the company, it becomes impossible to attract or keep good people. The HR employees who stay either started out or became unimaginative, passive, and very limited in their thinking and actions. They do not make a very strong group that could lead anyone at anything. In fact, HR is often a dumping ground for employees who no one wants. This helps perpetuate the mediocrity in the department, which is fine with everyone in the company who has influence or makes policy. It is ironic that although these same managers use the term "human resources" frequently, they are incapable of seeing their people as valuable resources for the company or HR as a department that could possibly help them.

The dysfunctional company that the turnaround CEO inherits has historically told its people through its actions that they are not a priority, and many of those people believe it. That acceptance by management and the employee base that people are not important undermines everything culturally that is needed to get a company believing that with good people who are well managed, great success will follow. They have been doing it so wrong for such a long time that the people program within the company is basically a wasteland. Except for a few isolated individuals and small groups of people who think differently and may actually be trying to do it right, the CEO will have little to draw from as he starts to remake the people culture in this bad company. The new CEO should assume that he is starting out with basically nothing other than an incumbent culture that does not understand or appreciate quality and opposes any form of change. There may be small pieces of a program that are potentially useable in the transformation, but they are an anomaly, are not widely embraced, and probably have only minimal value for the overall transformation. For all practical purposes, there are no processes, no standards, no training, or anything that is widely embraced and focuses on creating an environment for quality personnel. And there is no interest in any of this as well. The CEO will have to rebuild from scratch, starting with an impotent HR department that does not understand what its job should be. This will be a lot of change for a company that likes the status quo. The CEO will have to challenge everyone, especially senior leaders and managers who have to adopt a totally different perspective or leave the company.

The key initiative effort that follows utilizes the actions presented in chapter 9, but they are tailored to meet the specific requirements of a people transformation. The first subsection, "Make People First," not only creates the necessary sense of urgency but also addresses the cultural change that must take place and reaffirms the CEO's commitment to the transformation. "Find the Right Leader" constitutes building the guiding coalition with a new HR leader and his core team. Completing the coalition, aligning the company behind a great plan, and taking the plan forward is "The People Process," and "Never Let It Get Old" is the CEO staying involved and following through as well as the institutionalization of the program into the fabric of the company.

Make People First

The biggest obstacle the CEO faces is the company culture, which does not value people. Everything else is a by-product of this prevailing attitude. The CEO has to remake the culture of the company when it comes to how the company perceives its people and how those people perceive themselves. As is usually the case with broad transformation efforts, only the CEO can do this. This is a very important "stake in the ground" event, because it declares the old value system obsolete and defines a new direction for the company. The process of building a new company culture is covered in chapter 8, but this section specifically addresses the people part of the culture transformation. And this part is particularly important because if the company culture does not value people, anything else is just not enough.

The message from the CEO must be very specific and very to the point. It should be clear to all that this is not a negotiation but a declared new way forward. The new company culture will revolve around people because it is only possible to build a great company if that company has great people. The whole process of attracting highly competent people to the company, training those people, and helping them realize their potential at the company will become a core competency for all levels of management. Everyone will be involved and no one is exempt. There will be specific processes for personnel management, and everyone will learn them. HR will lead the effort and own the processes, and there will be zero tolerance for noncompliance. Everyone must get on board, especially senior leaders and managers. People are always interested in what they personally get out of a transformation. This is a great opportunity for the CEO to begin communicating to managers how they will personally benefit through better skill sets, having a more effective team under them, and a working environment and culture that are more amenable to success and the rewards that follow.

The CEO must go on to state that managers will be account-able for the quality and performance of their people. A significant part of their compensation will be tied to how well they build and manage their teams. Employees as well will have responsibilities within the process related to identifying and addressing their per-formance gaps. The people process will no longer be an orphan, since everyone will own a piece. HR will administer the processes, which means they will control how they are used and will be responsible for the content of those processes and any updates or modifications that are needed. Since they will have process expertise, HR will be expected to partner with managers on per-sonnel activities and be intimately aware of what each manager is doing in the personnel arena. HR will also be responsible for the employee training process, which will include having the appro-priate staff to support the function, working with management to determine training needs, and overseeing actual training and the evaluation of that training. The HR head and his staff will also have accountability for how well the people process functions and for the results. The focus will be on building a great team, addressing weak areas, and helping people improve and reach their potential. This will require a strong partnership between management and HR, and each group will be evaluated on how well that partnership works. The CEO will be expecting regular reports from all parties so that the process remains fresh and vital.

A key message within this people transformation is the level of effort required from managers for implementation of this new people program. Our experience tells us that in most bad compa-nies, managers spend much less than 5 percent of their time on their people. In many cases that number is practically zero. Bossidy and Charan, in their book *Execution*, tell us that in normal conditions, that number should be at least 20 percent and approaching 40 percent or more when there are big issues. In several interviews Jack Welch has said that he typically spent 50 percent of his time on his people. The challenge to managers here is that they must change their entire approach to their jobs. The days of doing what they want and what they are good at are over. Now they must do the whole job and learn how to do the parts that have always been a mystery to them. There are also some efficiency issues here concerning time management, and those are discussed in chapter 14. The specificity of this particular message will get everyone's attention. The CEO must hold the line on this requirement and be ready to deal with the naysayers who will say that such a requirement is unrealistic. In a bad company, that is probably true. The CEO must make the strong case that, in good companies, this is the standard and that managers who want to be on the team going forward will have to develop the skills and the attitudes to pass this hurdle.

For this cultural shift toward people to be truly legitimate, it must transcend the inner workings of the company and be seen and believed from all vantage points inside and outside the company. The people program must be as identifiable with the company as the company logo. The company should create people-related metrics that become part of the company lexicon and are discussed at virtually every meeting and at all operational reviews. The CEO should be regularly reporting progress on the people program internally and to the board of directors, Wall Street, and potential investors. Everyone who is paying attention should get the message that this CEO sees his people and what they are doing as every bit as important as revenue, earnings per share, and stock price. This makes it clear that in this company people are a real priority and it creates a commitment and expectation that the company management and everyone else has to respect.

Find the Right Leader

This new cultural shift for the company is all about great people doing great things. That applies to HR as well, since they will be such an integral part of this people effort. It is unrealistic and dangerous to expect mediocre people to do great things, so the HR team will require a rebuild. The HR head needs to be a game changer who commands respect and is clearly up to the challenge. The incumbent HR head is not that person and cannot become that person. Most efforts to get companies focused on employees fail because of a lack of commitment and/or inadequate leadership. Nothing will make a stronger statement about commitment and leadership than bringing in a very impressive person who is clearly capable of leading the charge.

Finding and hiring the right person for this position is really the first *acid test* for the company's commitment to top-quality people, and it is a difficult test, because the people infrastructure is not yet in place to help. The company must not only do a first-class job with the hiring process, but more importantly, it must produce the right person, which will be a first in both cases. Some top-notch outside professional help is definitely warranted here, and we advise against trying to economize or cut corners on this critical hire. The CEO should also be close to this effort so that everything goes smoothly, everyone stays on the plan, and the CEO gets what he wants in the position. The position demands a senior leader from a company with a people-centered culture who has experience leading change and operating successfully in adverse environments. This person could easily be a more seasoned and capable executive than anyone, other than the CEO who is in the company presently. The job description should be carefully crafted so that the right attributes are captured and appropriate candidates are attracted to the position and selected for interviews. It is vital that candidates can see the value of the position in the job specifications and that this is stressed again during the interview process. The candidates should hear directly from the CEO that this hire is of great strategic importance. The team charged with filling this position should work the process and interview candidates until everyone—including the CEO—is satisfied that they have found the right person. It is easy to get frustrated and rationalize the selection of a candidate who does not really measure up. This is a recipe for failure. The transformation of this company and the new culture discussed in chapter 8 are not going to happen successfully if the people piece is not totally put into place. That depends on the new HR leader doing the job, and an inferior candidate will not be able to rise to that standard.

The new head of HR must be an impressive person who is clearly up to the task. Even the Terrorists and the Rebels need to see this person as formidable and have some trepidation about their prospects of opposing him going forward. The CEO must introduce the new head of HR by restating the new company direction—which focuses on people, describing the role of the new HR head, and voicing his total support for the larger responsibilities of HR and its leader. It is crucial that the ownership of the new people transformation starts to flow from the CEO to the HR head, and that the HR head begins to take over the identity of the driving force behind the movement. This should not be seen as a complete handoff. The CEO should continuously demonstrate his interest and involvement, but the HR head should be closer to the work and accountable for making the effort successful. This move will broaden the transformation power base and give the HR head the credibility and influence that will be needed to rebuild HR and reform the people culture in the company.

It is very important that the CEO be accessible to the new HR leader and always ready to provide the necessary counsel, give a speech, have a sidebar conversation with a senior manager, or do whatever is necessary to keep the people transformation moving forward. It is crucial that everyone see the close partnership between the CEO and the new HR head—especially the new HR head. If the right person has been hired, the new HR head can be successful, provided the CEO does not get distracted or lose his nerve in the face of serious opposition. As previously stated in this book, the opposition will always attack the CEO's ownership of the transformation. That is the potential weak link in the chain. If the CEO and the new HR head can form a pact up front and stay together on their commitment, the people piece of the transformation can be successful, and that is a huge part of the program.

Before we move on to the actual development and implementation of the people transformation, we must address one very important challenge for the new HR head. As we have said, this people transformation is perhaps the most or one of the most vital parts of the company rebuild, but it is a tough concept for the rank and file to embrace. It strikes at the heart of their old value system and demands that they learn new skills and change the way they act and behave. Employees in bad companies are a pretty cynical lot and they will be looking for reasons to abandon this ship and get back into their comfort zone. Many may initially be impressed with the credentials of the new HR head—but that will not last long. They have to see positive action soon and then be continuously reassured that things are on the right track. They will lose interest quickly and assign "flavor of the week" status to the HR effort if there is a lull in the action or if big problems develop and roadblock the program. Once you have their attention, you cannot afford to lose it. The HR head is the key player in making sure there is no gap of inaction between all the promises and all that must take place. So having the HR head cruise into the company and spend the next six months getting to know people and building his team will not work. He has to start producing good things immediately. The question is how do you begin almost completely from scratch and make so much happen so quickly and have it all be effective? There is definitely a right way and wrong way. We will get into this very important but complex chain of events in the next section.

The People Process

As we said in the previous section, this transformation must come out of the gates quickly with virtually no time gap between the promises and the critical activities that must take place to initiate the program. The HR head must be a powerful leader, and that leadership is needed now more than ever. In chapters 4, 7, and 9 we talk at length about building a coalition of key people to make any change happen. This is particularly important here because the transformation strikes at the heart of the biggest dysfunction in the company: its people and how they are selected, managed, and valued. *Job one* for the HR head is to walk in the door with the core of his coalition in place so that these people can go to work immediately. If the HR head is the quality leader who is needed for this program, building a core team quickly should not be a major problem. Great leaders almost always bring some great people with them and get the rest based on their reputation. This core team must not just be good but real leaders and experts in making this transformation find success. These impact players will have a lot on their plate, and there is no room for error.

The two most important key first tasks that this core team must address right off are: 1) start fleshing out the basics of the transformation program, including the initial generic modules; and 2) take a hard look at the incumbent team and any other potential candidates for the coalition. Generically, most good people programs all contain about the same basic pieces. These consist of the following:

- Standards for positions—All key positions must have detailed job descriptions that define duties and basic responsibilities. The description should include required education and experience as well as any initial training needed to prepare the employee for the position. Another aspect here that is often ignored involves the personality traits that the job requires.

- Forecasting personnel needs—This is an HR-managed process, where managers determine their future manpower needs based on their expected future activity level and direction, and their evaluation of the people they have now. This process leads to building pipelines of the right people to fill those prospective positions and keep the company at maximum capability.

- Interviewing and hiring skill sets—This is another HR-managed process, where managers are trained to build capability at interviewing and hiring candidates for employment. The focus here is on identifying and properly evaluating prospective employees so that the best talent is brought into the company.

- Performance management—This is a process for managers to work with their employees to define expectations and realize maximum results. It involves goal-setting, performance feedback, metrics, mentoring, reviews, and rewards. The process components are standardized, but the details are customized to fit the position and the person.

- Career planning—This is a subset of performance management and forecasting personnel needs in that the manager and employee work together over the long term to build an experience base for the employee that addresses the needs of the manager and also prepares the employee for desired advancement.

- Gap analysis and training—This is another HR-managed activity where employees are evaluated for skill gaps relative to their job description and then provided with the necessary training and mentoring to close those gaps.

- Force-ranking of management—This is a process where managers rank those managers reporting to them based on overall performance. The results are used to determine rewards, promotions, training needs, and various corrective actions, including termination.

This list is pretty standard for the new HR core group. They have done this before and know what needs to be accomplished. They come from companies with great people cultures and they know what the basic processes are and how they work. They have all built out and activated these processes before and understand the requirements and most of the potential pitfalls. They also know what they do not know and how to ask the right questions. None of this is theoretical for them; they have lived in the environment that they are going to help create. They can jump right into developing the framework for the HR transformation program and getting that first task moving. The HR head needs to use what company information he has learned firsthand or from the CEO and others to focus the team on the highest priorities. This may involve a module that the company already partially understands and accepts in concept or an area that can be developed easily and will have a big impact on the company. The key here is to keep the program moving and showing positive benefits. Now is not the time to get bogged down in a very complex and controversial (at least to the rank and file) module that is difficult to craft and even harder to sell. Once the team has won some incremental victories and understands the company better, those tough modules can be addressed. *Task 1* is all about showing progress and gaining credibility.

An important side issue in Task 1 is getting the project scope right. When the driving team has great creative and productive capability, it is vitally important that they be directed and controlled by a plan that specifically sets the scope and defines the final outcome. A grandiose plan that is way beyond the ability of the company to assimilate could spell disaster, not only for this particular people transformation effort, but for the larger transformation program and the effectiveness of the CEO as well. The CEO and the HR head must initially work closely with the new HR core group to determine what are the critical gaps, what can be done to address those gaps, what are the key priority items, and how quickly the program can move and still be effective. The thing to remember here is that this is not a one-time event. All through the development and implementation of this program, the HR team needs to be reviewing the scope and the pace of the project and making adjustments based on what they have just learned. The incumbent team members who we will meet next can play a valuable role in these feedback reviews.

The second task of building the rest of the team is tricky and not nearly as straightforward. People are generally more complicated and unpredictable than processes and plans, but fortunately the HR head and his core team should know what they need. Since high expectations for the new HR team demand high standards for the team members, it is doubtful that very many of the existing HR team will make the grade. However, the transition demands that the HR team quickly develops a keen understanding of the company, develops a connection with its people, and builds credibility with those same people. That requires a capable incumbent contingency. These incumbents can come from the corporate HR team or from HR positions in the field or even from other positions outside but touching HR. The HR head and his core team must carefully scrutinize all these people and still maintain standards. There is no room for mediocre people in this effort. Although dysfunctional companies have few outstanding people, they do exist, and this team must find them. The final expanded coalition must be capable and very impressive, and most importantly, they all must be dedicated to the transformation and capable of working together as a team to make everything happen, and happen well. It must be clear to anyone paying attention that the new HR team is the strongest group of people in the company. This is not a preferred condition but an absolute requirement.

With the HR coalition now in place and the basic program framework under construction, what is important now is: 3) integrating the new members into the coalition and getting them involved in building out the rest of the program, and 4) growing the credibility of the HR team and its work product in the eyes of the rank-and-file employees. These two tasks are basically linked and can be thought of as one large task with one subtask naturally flowing into the next. In other words, you have to do Task 3 in the right fashion or you have little hope of getting Task 4 done. We will explain all of this in the next several paragraphs. This is vital to the success of the program, so pay attention.

As was discussed previously in this section, the HR team going into Task 3 will consist mainly of new people (the HR core group) with a minority of incumbents. The new people will have a background in good companies that value people and they will bring the basics of most of the new processes with them and will have built them out based on their experience. This is important because the program will need processes that have a proven track record. There is virtually no room for processes that may or may not be effective. Credibility is critical, and that comes with success and can leave just as quickly when there is failure. The incumbent team members have a vital role to play here. Although most of these new processes are foreign to them, they understand the critical issues and the underlying politics in the company and can really help the new core team customize their basic modules and direct the necessary emphasis and priorities where they are most needed. In some cases, there may exist a few best practices that have merit and can be integrated into a process. This gives the employee base some initial connection to the new program and shows the rank and file that this program is not just edicts but actually contains some dialogue. The incumbents are invaluable with this kind of work. They also have the background and perspective to make sure the program not only meets the needs of the company but will work best for the capabilities of the management that exists. The final product will not only be better but the required interaction within the team to get all of this done will solidify the team and make it more effective. Not only does the team need to see this but everyone on the outside will take notice as well. We will expand on this very crucial issue next.

At this point, we are at a key juncture in taking this transformation forward. This may be redundant, but let's review exactly where we are in the chain of events. The new core team is in place and is ably led by the new HR head. The core team has laid out the basic processes and added specific detail to the generic program, based on what they think is needed. The HR head and core team have scrutinized the incumbent HR team, as well as other possible internal candidates, and have brought in the best people available to fill out the rest of the team. And these incumbents are helping the new core team understand the company and make the necessary adjustments to the program. So far everything is fine, but this is where things can get off track if the team cannot make the transition from Task 3 of finalizing the program details and getting the new team members involved in the transformation to Task 4, where the rest of the company now needs to get invested in where this program must now take the company. The rest of this section focuses on how exactly that very important transition must proceed.

The first vital step in this process is making sure the HR team is totally together and has one identity. If this team cannot get together, then they have no chance at getting the rest of the company on board. One big mistake that people make on teams is creating internal caste systems. Here, we have the new core team and the incumbent players. During the final development of the program, those identifiers must go away. Initially the incumbents may feel like survivors who are lucky to still have jobs. The new core team members must treat them as peers and respect their perspectives and actively seek their advice. At the end of the creative process, all team members must feel like they played an important role and are just as valuable to the team as anyone else. So not only what is in the final program but how it got there is going to be crucial when it is time to take the program forward. Until the HR head feels he has created this kind of balance on the team, there is no point in taking the next step.

If Task 3 has been done correctly, you will have a good transformation plan in place, or at least a significant piece of it, that has been built by perhaps ten or twenty people (depends on the size of the company and the complexity of the transformation; in some smaller companies, the team could be fewer people) who are now behaving as a team. Now is the tough part. In Task 4 you have to take this new plan and not only communicate its contents but convince hundreds or perhaps thousands of people that it is right for them. If it were only that simple. As we have pointed out many times so far in this book, the people who fill the ranks in a dysfunctional company do not naturally respond well to change. In fact, they are experts at derailing change efforts. This transformation strikes right at the heart of what they value and how they act. Just telling them to do it will not get it done. This is a massive new step for them, and they have to be convinced by very credible people that this change is worth it. There is only one way to do this and it requires a lot of patient hard work by very capable people who are committed to making Task 4 work for everyone throughout the company. We will get into the specifics of Task 4 next.

As we said in chapter 4, this next task is about "expanding ownership." The HR team has to get a significant piece of the employee base intimately involved—actually, "invested" is a better word—in this transformation and then everything that must follow can be accomplished. So how do they get that done? We touch on it generically in chapter 4, but what follows here is the step-by-step process to make it work for this specific people transformation. While the nuts and bolts are important, what is critical here is changing peoples' values and how they view themselves and their fellow employees. Consequently, we have to operate at a more visceral level, and that requires a huge focus on human interaction. Communication is important, but turning that whole process into a deep personal conversation that inspires trust is what is needed, and that requires a lot more than just communication. The process must move beyond mere conversation and become a collaborative effort where participants from both sides get actively involved in understanding what is to be done, whether it makes sense, whether changes are needed, and what else must be done so that the transformation is as effective as it can be. There are three critical pieces to this program, and they are:

- Communication

- Interaction

- Modification

We will get into how each piece works next.

The CEO and the HR head can get Task 4 started by communicating to everyone that they are solidly behind the program, they will be involved, and that it will be successful. It reminds everyone that senior leadership is committed to this change effort and it makes sure once again that people understand where the company is headed. People need to hear many times what this trip is all about and why it is needed, so the CEO and HR head should revisit this topic many times throughout the transformation. However, this is only a piece of the overall communication effort. The more elusive piece involves the HR team and their need to create a connection with the rank and file. This can only be done by finding managers and other influential employees out in the field who have credibility with their peers and also have at least a partial understanding of what needs to be accomplished. The HR team should start working on *Day one* to find these people, then keep them up to date during development on what is transpiring, and then solicit their input on initial impressions of the program and what is missing or should be changed. As each piece of the program is crafted, HR team members should be getting input from these potential disciples out in the field. The incumbent HR team members can start the effort because they are known in the field and enjoy some trust, but they should be getting the new HR team members involved as well. This communication piece is all about building a beachhead of involved people within the employee ranks, and that can only happen if they are kept in the loop and asked for input. Cultivating a cadre of support, or at least potential support, out in the field during program development is absolutely vital to the effort. These people can do more to build interest within the ranks than anyone in the corporate office, even the HR incumbents, because: 1) people in the ranks know them better and see them every day, and 2) they, like their peers in the field, are on the receiving end of all the change and are going to be expected to use the new programs as well. All of this gives them natural credibility that no speech, pamphlet, or information docu-ment can duplicate. A smart HR team should know this and should be building connections out in the field as soon as they identify the right people out there who can help them move the program. If everyone does their job, word will get out in the field that this program has merit and the HR team is a group that can be trusted. The program is only ready to move onto the next phase when the HR head is very confident that there is an active area of support at least in a part of the company.

People often assume that the implementation phase of any program is focused on training people to use the program mod-ules and making sure they do things correctly. That is all true, but that is really only a small part of the process. We prefer to call this phase "interaction" rather than "implementation," because what is needed here is not a monologue or instruction but a dialogue. That is how you get your target audience invested in what you are trying to do: make them want to be owners—along with you—in what is being done. The safety case history in the *Key Initiatives* chapter (chapter 9) made it clear that the interaction piece is critical and that it must involve a lot of "face time" between the implementation team and the members of the target audience. How do we get that done? First of all, the HR team should carefully scrutinize their potential partners out in the field. There are plenty of Terrorists and Rebels around at this point who would love to disrupt the proceedings, and they are well versed at playing the role of "strong supporters." The team should start the interaction effort with partners who they believe are the most capable and receptive. The team needs some "early wins," plus they will learn a lot about the program and themselves as they take these initial interactions through their paces. The rollout should start with the simpler and more straightforward modules and then move on to more complex, and perhaps more controversial, modules as success is achieved and the target audience is better understood. Some areas of the company may have the ability and appetite to move fairly quickly through the program, while others may require a much slower pace. Rollout team members need to manage this rate of implementation very carefully by continuously getting feedback from their field partners about attitudes and issues. This is no time to take their collective eye off the ball.

All rollouts have to contend with difficult groups out in the field. These are the ones that take the most effort but can be the most rewarding when all is said and done—and they finally get on board. The difficult groups are usually the ones that have few, if any, field partners. Management is aloof, or even hostile, and the level of interest is initially at zero. Our view on how to deal with difficult groups is to open the lines of communication on *Day one* but not get bogged down in putting all your effort into trying to convert them. Senior team members should regularly interact with the management of these groups to keep them informed on program progress and to understand their issues, but the bulk of the team should be focusing on groups that are more receptive so the program can keep moving forward. As the rollout effort matures, these difficult groups become outliers and there is more pressure on them to come around. In some cases, managers actually "see the light" and get their group involved in the change. The key to the entire rollout process is to prioritize your targets carefully and not get stuck on an overly aggressive plan. The team and the program naturally get stronger with time, as they achieve incremental successes and build their capabilities through experience. So waiting out a tough target is not a bad thing.

The *modification phase* is really an extension of the *interaction phase*, and much of it is covered in the next section. The most important thing to remember about modifications during the interaction phase is that the conversation must always be a dialogue ....but standards must be maintained. What this means is that the program must be open to reasonable modification in the face of persuasive argument, but it cannot be watered down to nothing in order to be accepted by everyone. In that case, the Terrorists and Rebels win. The HR team has to carefully manage all of this, and team members will have some very difficult conversations with field people along the way. They may even lose some field partners in the process. The key here is knowing when a change is necessary and when it is not. Programs can be damaged beyond repair when people get into a modification frenzy and changes dominate the landscape. The initial program was built by some very capable people working together and continuously soliciting input from their field partners. Tweaks should be expected, but a complete renovation makes little sense and the team should make sure they are not maneuvered into this type of destructive activity.

On the nuts and bolts side, manager training is a key task during rollout. There is no room here for complacency. Team members should continuously monitor trainer capabilities, program content, and the responses of those being trained. If the training starts to slip, then the whole program will not be far behind. This is often a potential problem when rollouts mature and everyone gets busy doing something else. The first thing to suffer is the training. How managers respond to the training also needs to be closely scrutinized, because it will be the best indicator to the practiced eye of who is on board and who is trying to obstruct. The Rebels and Terrorists are always alive and active. There is no room here for bad behavior because everyone is watching. Discipline and corrective measures must be quick and decisive.

Another nuts-and-bolts item is that the HR team is responsible for quality, and they must closely monitor progress all the time. Even though every HR team member is seasoned and the program contains "tried and true" processes, things will go wrong. The team must have no tolerance for anything substandard and must make corrections responsively. Rationalizing substandard behaviors or results has no place here. Everyone expects this team to be on top of everything and they must live up to that expectation. If they demand the best from themselves, then it is not unreasonable to demand the same of the company, its managers, and its employees. That is how rollouts start out successful and stay that way.

Big transformation programs in bad companies can take several years to be completed because the rank and file can digest just so much in the way of new processes and new practices. They will get better with time, but initially this is all very new and frightening to them, and they are not good at dealing with change. The HR team may be very capable, but the weak link here is everyone else in the company. Challenging them is good—but pushing them beyond their collective capability will not work. Many well-intentioned transformation plans fail at this very point because the program is moving too fast for the target audience. The first few pieces of the program will set the tone for what will follow. If the initial modules are well crafted and strongly supported by the HR team, then there is a very good chance that most of the managers and the employee base will do their part and start to get engaged. Conversely, if the initial program modules are mired in chaos, then this only empowers the opposition, and the program could get into serious trouble. Needless to say, the HR team must carefully plan everything related to the initial implementation and leave nothing to chance. The program must be well orchestrated, and the management and support must be perfectly tuned to the needs of the managers and their people. The HR head must make sure this happens—because the program depends on it.

Never Let It Get Old

Like any key transformation program, this *people effort* is not something you do and then forget about. It will require continuous attention as long as there is one manager and one employee in the company. The natural forces in the company will destroy this program if given the chance. The HR head needs to be very visible when it comes to the people program. He should be connecting with senior managers throughout the company on a regular basis, monitoring performance, calling attention to examples of outstanding results, and taking the lead on any major issues that arise. The HR head should be keeping everyone up to date on company performance of the program and monitoring, either directly or indirectly, every manager's performance. The HR team should be an appendage of their leader, working with all levels of management to keep the program running effectively and helping managers get the most out of the processes. Early on, it is critical that processes work and that everyone becomes proficient with the parts of the processes that they must use. Issues should be dealt with quickly and program adjustments developed and implemented when necessary. Once the HR team feels that managers have the knowledge base to run the processes, the focus needs to be on compliance. This is a very important step in the overall people effort, because the Terrorists and the Rebels know that they can sabotage the program if they can get managers to marginalize compliance. The HR team must maintain standards and have no toleration for noncompliance. They need to make it clear that they are there to help, but they will provide no waivers; everyone must perform. Persistent noncompliance should be addressed rapidly and consistently and should involve the appropriate levels of authority. Everyone should get the message that they are expected to do their part of the program to the best of their abilities, and to do anything less is putting their career in jeopardy.

The CEO should continue to champion the program and keep sending the message that the company is focused on quality people doing quality work. It should be clear to everyone that the CEO sees the people program as a key piece of the company strategy and a way to differentiate the company from its competition. The CEO should keep the message and the program fresh by challenging everyone, and especially the HR team, to find ways to make the program better. This should not be a wish but an expectation. Key HR people should be charged with looking for best practices outside the company and mining the rank and file for ways to improve what they are actively using. Keeping people first has to be a living program and a key value within the company. It is up to the CEO to make sure everyone in the company not only gets that message but believes it and lives it.

As was pointed out in the previous paragraph, part of keeping the program fresh is making needed changes and improvements. This is part of the modification phase. Here again, how you make those changes is just as important as what those changes entail. As good programs mature, more and more of the ownership moves to the users out in the field. A smart HR head will leverage that movement by getting field partners involved and perhaps leading efforts to make the needed changes and improvements. These people are closer to the actual work and probably understand better than anyone what needs to be done. Plus, it creates great optics when field personnel initiate and then play a big role in implementing the needed changes to a well-functioning people program.

What You Can Expect

The big mistake here is to assume that this will all go smoothly. That is an unrealistic expectation. The company is still a bad company with a lot of bad people, and failure has been the norm for a long time. Plus, there are still a number of people "in the weeds," dedicated to making the transformation fail. Since this people piece is a key initial part of the overall transformation, it is not feasible to wait for conditions to be perfect. Leadership has to bite the bullet and take on the company, with all its warts and imperfections. Many things will initially go wrong, people will push back, and there may be, at times, more bad news than good news. What is critical here is that the CEO and team, especially the HR head, not lose heart but continue to drive the project by staying committed and staying involved. The rank and file, as well as the Terrorists and Rebels who are still around, cannot see you sweat here—and failure is not an option and should not even be on the table. The CEO must keep messaging that the overall plan is still good and that making adjustments and dealing with mistakes is normal. The rank and file are used to leadership just wilting in the face of problems and initial failure, and they will be looking for that now. This cannot happen or the whole transformation will die on the spot. Issues need to be tackled immediately by the right people with the right resources. One special team should be looking at the whole program and all its pieces on a continuous basis to make sure all the pieces fit even after necessary changes are made. A common mistake often made with such programs is when so many changes are created on a "one-off" basis that the whole program no longer makes sense. Should that happen, it could be fatal to the whole effort. Another critical necessity is understanding why modules that fail are failing. Are the modules faulty or is the problem with the people employing the modules? If the problem is people, then is it a training issue, a personal capabilities issue, or just plain sabotage? Getting to the right answer and then making the appropriate fix is vital. This all requires the driving team to stay on point, stay connected to the users, understand what is happening, and be responsive and on target concerning the needed adjustments.

The CEO and the transformation team need to understand all of this in advance and not be caught off balance when things get complicated. It is impossible to preplan everything and not have to scramble, but generic contingency plans must be in place and teams ready to implement them so that the scrambling can be minimized. The key piece here is understanding the commitment required and making it happen every day no matter the depth or breadth of the issues. That is what is needed for a successful people transformation.

Final Comments

The people program is a key initial piece to the overall transformation. Along with safety and performance management, this initiative solidifies the new culture. As the people get better, either through training and mentoring or through replacement, the company gets better. Managers on down to the rank and file are not only more capable but they have better values and behaviors. This builds the foundation for all that follows on this organization's journey toward becoming a good company.

Chapter 12

Performance Management

Chuck: This is the third pillar key initiative, along with safety and HR, that is required to get the transforming company on a solid footing. We built a safety culture to protect our employees in chapter 10, and in chapter 11 we developed the capability to identify our personnel needs, hire the right people, and prepare them for the work ahead. In this chapter, we will talk about how these employees should be managed so that they can most effectively do their jobs and build their careers.

The previous chapter on the HR Function made reference to *Performance Management* as one of the building blocks in any good people program. In fact, this topic is so important that it deserves its own chapter. In the previous chapter, much of the focus was on the roles of the CEO, HR head, and the HR team in building and implementing the people transformation. Managers at various levels and their roles were mentioned, but they were not the main thrust of that chapter. In this chapter the focus is totally on the manager and the subordinate interface, independent of their level in the company, and what goes on at this grassroots level, so that both parties can benefit from the interaction—the work gets done properly, success is attained, and careers are built. The paradigm that should exist in this relationship between "boss" and "employee" is pretty much the same no matter whether it is between a CEO and an SVP or a section manager and a staff member. The responsibilities and activities are basically the same and just as critical, and no one should get a waiver on doing their part. Performance management by itself cannot totally transform a company, but it is a very necessary practice and companies cannot find success without it.

Why Performance Must Be Managed

If you are a small business owner and have only one employee other than yourself, you need some form of performance management to get value out of that employee's efforts. You had an idea in mind of what you wanted someone to do for you before you actually hired that person. Hopefully you explained what the position involved and what you wanted that person to do on the job during the interview. Ideally, the person you hired had the right experience and pretty much knew how to do the job or at least was capable of being trained. That is all a good start, but the process cannot end there. Even if the person has the right experience, this particular job will have some differences. It is your work environment and you have to articulate how you want the work done to properly benefit your company. You owe it to yourself and to the new employee to make your expectations clear during the interview and again on day one. If the employee needs some training or at least a demonstration, this is a great time to deal with all of that. Once the employee is ready to do the job, you owe it to yourself and to the employee to regularly monitor the work, give feedback as needed, and do some retraining if necessary. To get the most out of the process, you will also ask the employee for any feedback he may have about how the job could be more effective or efficient, and you will also give praise and preferably rewards when the performance is really outstanding. If the employee is struggling because of skill or work habit problems or demonstrates an attitude that is detrimental to the job, it is your responsibility to bring the issue to the attention of the employee, hear his perspective so that you fully understand the situation, and then develop a program for him to hopefully resolve the issue over a specific period of time. If the employee is able to resolve the issue with your input and assistance, then you should provide the appropriate feedback that you are now satisfied with his performance, and life returns to normal. If the employee is unable, for whatever reason, to successfully deal with the issue, then you must explain your position to the employee and terminate his employment. In either case it is critical that you communicate the situation to the employee, understand his position, make an honest effort to find a solution, and be fair and consistent in your treatment of the employee. In this way, you make it possible for the employee to maximize his opportunity for attaining success at the company. This is good for you and good for the employee.

Your initial response to all of this is probably that this is just Management 101. Common sense would tell you that this is the only way to do it. If that is true, then why is it basically nonexistent in the bad company? Even in good companies it can get sloppy and inconsistent. There are basically two reasons and they are both rooted in the culture of the company. The first reason is that managers are really managers in name only and do not know how to manage their people and their work. They have never been trained to manage their people, and there are no examples around from which to learn. Their boss does not manage them, so the bad behavior trickles down through the organization. Effectively managing people is not valued in the bad company, so virtually no one focuses on it or attaches any importance to it. Managers know they can fake it and get away with that kind of behavior because everyone is doing it that way. There are a million other things they would rather do that are more interesting to them and not nearly as mysterious as managing people.

The second reason is that performance management involves a lot of work and it is hard to do it right. You have to make the effort to understand what your people do, how they do it, and how they should be doing it. You have to spend enough time with each person to understand their skill sets, their gaps, their tendencies, and their aspirations. If you have ten direct reports, then you have to make the effort to know firsthand ten different positions and the people responsible for those positions. You cannot do that just sitting in your office. It takes initiative to dig into those positions, ask questions, evaluate responses, and form opinions about what is really happening. And you cannot just do it once and be done with it. It is a continuous process to keep your information base fresh. Goal-setting requires that you understand where each person is in their development, what they are capable of, and what the company really needs from them. Then you have to measure performance and not only be able to spot issues but get involved in developing the solutions on how to fix them. Then you have to have a lot of conversations with each of your employees about all of this, and some of the things you have to say they may not like to hear. None of this is much fun, but in the good company it is an absolute necessity. Good companies know that making a commitment to performance management and holding everyone accountable to that commitment is how companies find success. That is part of their culture. In the bad company, they keep their people at arm's-length and pretty much on automatic pilot, minimize or eliminate the work on goals, and keep reviews and feedback pretty general and not to the point. That way no one has to work very hard, everyone is happy, and life goes on as usual. That is what they value. Unfortunately, it is a recipe for failure.

In the company example just discussed where the owner had only one employee, it was still necessary to have a process so the owner could be assured that he got the desired result from the employee's efforts. In a larger company, the process is basically the same but more complex and even more important. The truism that good companies understand is that consistently good results are only possible through management. When people are put on automatic pilot, virtually anything can happen. If, as a manager, you plan to rely on people's good intentions to get the results you want, then you can plan on being disappointed consistently. Employees, independent of their skill sets and attitudes, have to be told what is expected, then shown how to do their jobs, and then given feedback along the way. As a manager, you can now have some confidence in expecting that your employees are functioning in the prescribed way and that you will get the results you want. If you are doing anything less than this, then you are not really managing but only hoping. And hoping has no impact on performance.

What It Looks Like

Before we get into the nuts and bolts of performance management, it is important to talk about management style. People have different personalities, and while some managers are warm and fuzzy with their people, others are naturally more aloof. Independent of how people are wired, it is crucial that managers take the time to know their employees and to manage them in a face-to-face manner that makes room for dialogue. If a manager cannot do that, then that person has no business managing anyone. Bad managers often just ignore their employees and then yell at them when things go badly. The manager is never at fault; it is always the employee who is to blame. Another common approach is to manage by edict. All interaction is one way, and standard marching orders would be: "Make those numbers I gave you." While *people management* is not a democracy (decisions are not made by voting), it is not a dictatorship, either. The manager is responsible for getting certain things done and has transferred some of that responsibility to the direct report. All that requires a lot of dialogue between the manager and the employee if things are going to go well. What is expected? Is it feasible, given business conditions and available resources? How can it be achieved, and are there special needs in terms of people and resources? Is there a better approach? Good managers ask a lot of questions rather than issue edicts. In fact, one senior manager in a large service company was called Twenty Questions behind his back because he could usually get to the bottom of even complex issues by asking twenty questions or less. When a manager stops issuing edicts and starts asking questions, it is amazing how much better prepared direct reports are and how much they already know. The key message here is that managers have to get involved with their people and the work they are doing. That does not mean micromanagement, but it does mean knowing what is going on. Give your people room to offer suggestions, make decisions, and use their judgment. If you spend enough time with them, they will know what the rational boundaries are and when they are no longer on firm ground and need help. If you want to get the most out of your people, you need to start talking to them.

We referenced a book, *Execution*, by Bossidy and Charan in the opening chapter and talked about it again throughout this book. The rest of the title of that book is: *The Discipline of Getting Things Done*, and that really says it all. If you want to consistently get things done in a company, you need a disciplined process and then you have to continuously work that process. The book is primarily focused on the management interactions between the CEO and the next level down, but much of what is in that book can be applied to about any level in the organization. The authors talk about seven essential behaviors:

- Know your people and your business

- Insist on realism

- Set clear goals and priorities

- Follow through

- Reward the doers

- Expand people's capabilities

- Know yourself

These behaviors are totally consistent with the good company culture, described in chapter 8, and the people-management behaviors, discussed in chapter 11, on the HR Function. This should not be surprising. There is really only one basic way to do it right. We will use these behaviors to frame the performance-management process later in this chapter. By contrast, it should be obvious to the reader at this point in our book that bad companies are 180 degrees out on all of the preceding behaviors. A rather ironic example of this involves the president of a large service corporation who had his senior managers in for a review session.

In the room were a number of boxes of books. As the books were being passed out to the participants, the president remarked that this book, titled *Execution*, was very good and that the participants should read it. The book was never mentioned again at the meeting or at any future meetings. One participant did follow up, on his own, with most of the other participants and did inquire if they'd read the book. In all cases, the answer was negative, and in a few cases, the person did not know where the book was now located. This president either had not read the book or had forgotten the fourth behavior, "follow through." It was not surprising that he was getting the quality of performance from his people that he deserved.

When it comes to managing people and their work, there are two basic truisms that the reader must keep in mind. The first is that results are only controlled through management. We have touched on this already, and the Bossidy and Charan book makes a great case for it, but it is crucial that we communicate the basics here. As a manager, you have to have a clear view for yourself of *what* you expect and *what* success looks like. And then you have to communicate that to your employees right up front, so they know what is expected. It is amazing how many managers give little thought to *what* they want, and even if they have some idea, they often fail to share it with their people. It is hard to get people moving in the right direction with purpose if no one bothers to discuss the "what" with them. And if the "what" is very elusive in the bad company, then the "how" virtually does not exist. Bossidy and Charan devote much of their book to building the "how," which involves the details of how the "what" is to be achieved. Most bad managers, if they do anything, just throw numbers or desired outcomes at their people and leave it at that. Where the desired outcomes are generated is anyone's guess. They probably come from the accounting department, but they have nothing to do with reality. The "how" is rooted in reality and should be developed in concert with the "what." In fact, a "what" without a "how" has no validity and should never happen. It is an injustice to your direct reports to hand them a "what" without first working with them to build the details of the "how." When managers work with their people to build the "what" and the "how" for the coming year, and then support that initial effort with mentoring and reviews, help them deal with issues, and then reward them when they do well, then you have a management system that can help people achieve goals that can move the company forward.

The second truism is that virtually all people need a quality system to be successful. A very small percentage of people may be able to figure it all out and be successful totally on their own, but that is very rare and you cannot rely on it. Successful employees typically have a close working relationship with their boss, who not only functions as the manager but is basically a coach, and that coach provides and administers the quality system the employee needs to excel. As we have said previously, when a manager moves to the mentor/coach level and does a complete job on his part of the performance-management process, the required effort and time commitment is substantial but absolutely critical for success. The manager can start out by adopting the seven behaviors presented in the preceding. Take time to understand the business you are in and take even more time to know your employees, their skill sets, personalities, gaps, aspirations and, most importantly, the jobs they are doing. You should know how they are going to react to certain situations before they do. In your dealings with your team, throw out all of the fantasyland views and sales talk and demand realism on everything. Challenge people when they lapse back into these old habits. You can only manage effectively with the truth, and make sure everyone understands that. Set clear goals and priorities for everyone, including yourself, and make the process a dialogue and put them out there for everyone to see. There is nothing secretive about what people are doing. And unlike that president of the service company, follow through on everything. If you ignore it, then it will not happen. Build a reputation for being tenacious about getting things done. Ask a lot of questions all the time. You will be surprised how much more your people start knowing when you are always asking them questions. Celebrate and reward success and make sure the right people are recognized. Nothing is more de-energizing than a boss who hogs all the glory and the credit or rewards the wrong people. Challenge people to get better. Continuously put new skills in front of them and get involved in their short-term and long-term personal goals. Help them be as successful as they can be. And last, (remember chapter 3) be honest and fair with everyone, including yourself. Your people deserve the truth, so give it to them no matter how difficult or painful it may be at times. If you are consistent and forthright when you deal with others, you will be surprised how the practice rubs off on the people around you.

Once you have the basic behaviors down, then it is a matter of doing the process, which generally involves the following activities:

1. <u>Pick the right person</u>—Make sure the employee has the basic ability to do the job. No amount of management and coaching is going to lead to success if the employee cannot meet the intellectual and emotional requirements of the job. Good managers put a lot of effort into finding the right person. They know that good people put in the right position will benefit the company more, and it is easier and more effective to manage good people. While it takes less work to hire just anyone, the costs in time, output, and actual dollars of dealing with the resulting substandard performance and the higher turnover that follows are significant. Recruiting and interviewing are the needed skills to do this job correctly. Partnering with HR, as discussed in chapter 11, is an absolute necessity to build the candidate pipeline and receive help in honing those interviewing skills.

2. <u>Training and teaching</u>—Provide the basic training for the job so that the employee knows firsthand what must be done. Determine any specific gaps and provide the appropriate specialized training to address those gaps. It is your responsibility to prepare the employee for the job and to give that person all the necessary aids and support to be successful. The close interaction of teaching and training at this stage will also give you a much better insight into the employee and his capabilities and build the foundation

for the strong manager/employee relationship that must follow. Be aware that training does not stop at the hire stage. Be open to providing additional training for your employee whenever skills need to be refreshed or additional capabilities are required.

3. Goal-Setting—We talked about building the "what" and "how" earlier in this chapter, because these are such critical areas that, if done correctly, elevate a performance-management program well above the norm. Effective managers do not just throw numbers at people. Goal-setting is a team exercise and good managers work with their direct reports to build goals that both parties believe are realistic but also contribute to the company's success. As pointed out earlier, people management is not a democratic process. People do not vote. But it is not a dictatorship either. Sometimes employees cannot see the opportunities that are available or may not be as aggressive as they possibly should be. It is up to the manager to walk them through what is possible and help them see what can be done. In the end, the direct report must feel that they have participated in the process and that they helped build and now own the goals. That only happens when the goals are born out of a dialogue rather than out of an edict.

4. Providing feedback and support—As manager and coach, you share a responsibility with the employees to see that they master the jobs they do. That means paying close attention to their performance and providing feedback on a continuous basis. This feedback has various components and includes much more than formal reviews every quarter or so. A good coach is paying attention on a regular basis and giving guidance when it is needed. The employees must feel that their performance is their responsibility, but it helps their development if they know they have a mentor who is interested and close at hand. More structured reviews should take place on a regular basis and

should involve an agenda that both parties understand and that requires both of them to assemble information in advance. Good topics include progress on goals, changed conditions, additional skill set development, any training that is needed, the status of any unresolved issues or any new issues, and possible career moves in the future. Good reviews are honest and truthful conversations between two people who respect and trust each other and value what they are doing within the process. As the manager, you have to take the lead on this and set the example.

5. <u>Maintaining accountability</u>—This is an extension of the feedback and support piece, but it is vitally important to the process and deserves its own place. As manager and coach, you set standards for yourself and for your people. That is what a quality process is all about—meeting and maintaining standards of behavior and performance. This responsibility has three components. The first one is that you must articulate those standards to your employees initially, and then again whenever it is required. It is key that your people understand exactly what you expect and that it is always foremost in their minds. The second component is that you then monitor activity in your part of the company to make sure that employee behaviors, actions, and the results that follow do indeed meet your standards. The third component involves following up and making corrections when standards are not being met. Many managers fail to realize that without this third component, there is no accountability and that doing nothing about a lapse in standards basically makes those standards nonexistent. Good managers move quickly on noncompliance. Once they have all the facts, they deal with it directly. This is not a responsibility that should be delegated. You have to be fair, consistent, and specific as you communicate to the appropriate parties what your understanding of the infraction is and how you will deal with it. The company and your employees deserve nothing less.

6. <u>Communicating success</u>—Not only do you have to act when things do not go well; you also have to act when there is success. It is amazing how many managers miss these opportunities to appropriately reward successful employees and communicate how much they are valued. Managers will pass out bonus checks as if they are no big deal and not worthy of any discussion. Success and rewards are a big deal. That is what performance management is there to do, so it makes sense to stop and celebrate that success when it actually happens. Rewards should be given out in a scheduled review meeting. The manager should express his appreciation, give his view on why things went so well, and solicit the employee's perspective. There should be additional discussion on how they can build on this success. The priority is to make the employee feel valued and to understand that, by doing their part in the performance-management process, they have found success.

These are not complex activities, but they involve a lot of work and you have to spend a lot of time dealing with people. It is definitely a discipline of getting things done. But if you religiously work this process and stay true to it, you will build a team of good employees and you will be successful. So why is it that so many managers do very little of this, if any of it, and rarely find success with it? We will get into that and how to fix it in the next section.

Building a Success Plan for Managers

A turnaround CEO came into a bad company in trouble and stated that he was going to change the culture of the company. In his opinion, the managers were not professional, had no respect for operations, and had no idea how they should handle their employees. He demanded that managers prepare goals for their direct reports and conduct regular review sessions on performance. Unfortunately, the directive gave no more detail. To make matters worse, the CEO was lazy in his goal-setting activities, and review sessions with him were usually conversations about hobbies and vacations. Needless to say, this program died a slow death and eventually all but disappeared. The message here is very simple: had the CEO been very specific about what he wanted, not exempted himself from the process, and then followed through to make sure everyone did their part, the program could have met with success. This CEO, like many managers, was lazy, was not committed, did not want to do the work, and possibly did not really know what to do. He was hoping for some form of miracle. That is a recipe for failure.

In bad companies the lack of effective performance management is due to the poor leaders at the top and the managers who work for those leaders. As was discussed in chapter 11, Bossidy and Charan state that managers should be spending as much as 40 percent and usually 20 percent or more of their time selecting, evaluating, and developing their people. Leaders and managers in bad companies do not know that. In the typical bad company, that number is probably much less than 5 percent (less than two hours per week), and in most cases the manager is dreading having to devote any time to his people. The reason for this huge gap between what good managers do and what bad managers do is that bad managers and their bosses do not know how to do it, have no history with it, and since they are not good at it, they do not want to do it. You end up with a culture where managers do not know how to manage and they and their bosses are in denial about it. That way it works for everyone. The only way to change this is to start at the top. Performance management is a trickle-down process. If your boss does it with you, then you get the message that you had better be doing it with your people. The turnaround CEO had the right idea, but he failed on the execution part. The CEO has to initiate the process with his direct reports and then demand that those people properly pass it on. The basic assumption has to be that this is all new for everyone and you have to start from scratch and do it right. That means going through the entire performance-management process with direct reports and sticking to every detail. There can be some training sessions with HR involved, but the one-on-one time with the CEO is necessary for several reasons. First, it has to be clear that this is very important and that it is going to happen, and the direct report needs to become very proficient at all the components of the process. Second, there may be pushback and some direct reports who will have none of it, so the CEO needs to be front and center to deal personally with those issues. If the CEO sets the example and stays the course, this commitment to performance management can be successful. It will take a lot of work and time, and getting people moving in the right direction, but it cannot be done any other way. Now let's get into the details.

The key to effective performance management is in the details. As we have seen, bad managers do not do any of it. Mediocre managers try, but the detail is lacking, so there are too many gaps in plans and commitments. No one really knows what is happening and the process unravels. You have to have the details, which means having the discipline to do all of it correctly and completely. That starts with establishing a mutual agreement on expectations, which again is all about the "what" and the "how." The CEO and the direct report (and later the direct report and the next level down) must define the state of things at present in every area that the direct report covers. They must agree on what is working, what has issues, and what are the existing numbers. Then they must build a plan for what is going to happen going forward. That plan will include what will stay the same and what will change, what will be added, and how those changes and additions are going to take place. The plan must include timelines, key activities, persons responsible, critical resource needs, and the numbers that result. There must be discussions about the direct report's team, including open positions, needed replacements or upgrades, skill gaps, people with issues, and new talent needs. The final discussion is about the direct report's development—including skill gaps, style issues, and specific plans for self-improvement. Although the direct report is responsible for building the detail in all the plans, the CEO should be heavily involved in more than just a review capacity. The CEO must spend enough time to really understand the issues and to be confident that the plans are on the mark. When the CEO gets deep into the process, it will be clear to the direct report that this is serious business and the CEO is paying attention to all of it. This approach also creates many good opportunities for real-time mentoring on specific issues. In the end, the CEO and the direct report should both have a very clear view of the state of things and the direct report's plans to make it happen going forward. No one should be having any vague thoughts at this point in the process. The direct report's program is basically a contract, so it all should be very specific for everyone.

After putting so much effort into the plans, it is critical that people do not get lazy on the review part of the process. The reviews and feedback are just as important because they keep the plans current as well as effective and help the direct report be successful. Like the original development of the plans, the review process must be as complete and focused on the details so that everything is covered and nothing slips through. Formal reviews should take place at least quarterly. In some cases, more frequency may be justified if there are difficult or key projects that require more attention. In addition, the CEO should continuously ask for updates and give informal feedback when appropriate. The key is to have enough dialogue between the CEO and direct report to keep issues fresh and projects moving without the process being more burden than benefit. In the formal reviews, the CEO should establish the program up front and stick to it so everyone knows what to expect. The process is working when the direct report comes into the review already knowing the agenda and most of the outcomes. Reviews are a great opportunity not only to communicate in both directions but to share ideas, address issues, teach, and establish accountability. The CEO must stay true to the process and be consistent, because not only is the direct report learning how to manage and execute but that information is being taken down the ladder to the next several layers of the organizational chart. How to conduct reviews, how to deal with employees, how to address performance issues and fix problems, and how to achieve excellence and reward it are being established for the whole company in those review sessions between the CEO and his direct reports. The litmus test for both the CEO and the direct report is that they both walk out of the review with no lingering questions about any detail on the status of any part of the program. They may have concerns, but those will have been communicated during the review. In bad companies, they like to keep reviews, if indeed they happen, as vague and murky as possible. In good companies, reviews are a time for clarity and the truth, and both parties have a responsibility for that.

Maintaining the Process

If the CEO does his part correctly and makes sure the practice is passed along, the company can internally develop a skill set for performance management and it becomes part of who they are. The end product is a management team at all levels that is proficient at performance management. By-products include better employees who are better managed, better operational results for the company, and a company reputation for doing things well, which yields all sorts of benefits related to business opportunities and higher interest in the company as a good place to work and a good place to invest. These are all great things, but they will not happen—or they will not last long—if the process is not adequately supported. Like everything else in the company transformation, you cannot just do it once and then put it on automatic pilot. The CEO must keep doing his part, which is properly managing the next level and then continuing to pay attention to the overall process with messaging, making it a focus area on goals, and being persistent in monitoring how well the process is working downstream. HR must stay involved with training modules, improvements to the program, and working directly with managers to hone skills and fix problems. Each manager must be personally committed to doing the process well and passing it on intact to his direct reports. Furthermore, each manager must be dedicated to building a rapport with his people and turning them into good employees and managers and helping them do their best and build great careers in the company. Everyone within the process must keep looking for ways to improve it and make it more effective. But most importantly, everyone involved in performance management must stay true to the process every day and never

let their standards slip. If that all happens, performance management will continue to be a core competency of the company going
forward well into the future.

SECTION IV

ANCHOR THE CULTURE

Chapter 13

Building and Supporting Processes

Chuck: In the two previous book sections, *Overcoming Dysfunction* and *The Basic Building Blocks,* we described the change process, how dysfunction can be fixed, and the crucial role the CEO has in leading the necessary transformation. We talked about building the team, developing a new company culture that has the correct values for success, and we got into key initiatives that are the building blocks employed to transform the company, the big three being: safety, the people process, and performance management. This last section of the book, *Anchoring the Culture*, focuses on what is needed to keep things going in the right direction. This particular chapter on processes talks about how good companies do their work. This is critical, because if managers and their people do not know how to work effectively, the transformation is going to be stopped in its tracks.

Back in chapter 2 we talked about the key characteristics of failure in the bad company. One of those characteristics is the inability of bad company leadership and management to understand, like, and employ strategy and processes in the normal operation of their company. These people have no successful history with strategy and processes, have little or no idea how they work, and see no value in them. Of course, in the good company the opposite is true. Executives in good-to-great companies are experts at strategy and processes or they would not be there. Creating a capability for strategy is a key initiative, and that topic is covered in chapter 9. As was pointed out in that chapter, key initiatives touch and affect practically everyone in the organization. They are a company-wide effort and the generic methods required to make that happen are laid out in chapter 9. Some processes are company-wide. In fact, being strategic does involve a company-wide process, and those types of processes do fall in the key initiative category. What we want to talk about in this chapter are the majority of processes that are not company-wide but affect only a segment of the company. The fact that they are in place and being used has a beneficial effect on the whole company, but the people who touch them and use them are only part of the company. That fact has a big impact on how you build and employ these more localized processes.

This chapter on processes is another component in the building of the basic infrastructure that all good companies have and all dysfunctional companies try to avoid. Just like the company culture, you can learn a lot about a company by taking a hard look at how the people collectively do their work on an everyday basis. Is it organized and efficient? Are people plugged into what is happening or is everyone just doing it their way? Is there strong communication both vertically and horizontally, or does everyone work in a silo? Is overall performance consistent or erratic, and does the work product have quality or does anyone care? How these, and many more questions like them, can be answered is determined by the presence or absence of processes in the company, and the effectiveness of those processes. In a nutshell, good companies have good processes, and great companies have great processes that keep getting better all the time. Great companies put a lot of time and thought into the care and feeding of their processes. They would not have it any other way. In contrast, bad companies never met a process that they liked. They usually do not have them. But if they have them, they do not use them or at least do not use them correctly. In this chapter, we will explain what processes are and what they do. We will review why bad companies do not like them and have no success with them, and then we will go into the right way to build processes and then keep them fresh and vital.

What Are Processes?

Any work activity that is repeatable lends itself to being controlled by a process. Basically, a process is the recipe or set of instructions for doing that work activity in a good way and, hopefully, in the best way possible. For example, holding a meeting can be and should be managed by a process. As we have already seen in this book, bad companies have a lot of meetings. In most cases, there is no agenda, or if there is one, it is usually not followed. Who attends is also fairly random. No one takes notes and there are no written findings. People leave the meetings with nothing resolved. A process for a meeting would require the initiator to make a clear statement about the purpose of the meeting, who should attend and why, and what the expected outcome would be. The initiator would also prepare and distribute in advance for comment the agenda and meeting schedule. Any prework assignments would also be made in advance. The meeting would start on time and stay on agenda unless the initiator and attendees agree that new items needed to be discussed. Someone would be assigned to take notes and the content of those notes would be part of the meeting discussion. At the end of the meeting the initiator would state the findings and conclusions and offer his assessment as to whether the meeting accomplished its intended goals. He would also ask for comments concerning any additions or changes. Once there is agreement on the meeting notes, as well as the findings and conclusions, someone at the meeting would be assigned to document and distribute them. Any post-meeting work assignments would also be made at the end of the meeting and the need for a follow-up meeting and the details of that meeting would also be discussed. Now, to some people this may appear to be overly complicated. But in fact it is just a clear set of instructions on how to set up and run a meeting that is productive. Once people learn the process and follow it a few times, it becomes second nature; and all of a sudden, a new group of people knows how to run a meeting that is not a waste of time. That is what processes do. They show you the *right way* to do things.

For processes to be really useful, they should have enough detail so that there is no confusion about what should be done. And furthermore, it is important to eliminate any avenues for shortcuts without violating the process. The process should touch everyone who is involved, from the person who actually performs the activity to anyone who that person interfaces with, as well as those who are in a supervisory role. If people have questions about roles or tasks after reading the process, then something is missing. By the same token, processes should not carry any excess baggage. If a sentence or a section is not really needed or is redundant, it should be eliminated. Efficiency and completeness are the key words in building a good process.

Not all jobs lend themselves to processes. For example, the CEO job cannot be defined by a process or a collection of processes. As we saw in chapter 6, that job involves a lot more than performing repeatable activities. For the CEO, as well as most executive level leaders, every day is different as new issues arise both inside and outside the company, and each issue demands its own unique treatment. However, even these people will be touched by processes and often perform tasks or participate in activities that are covered by a process or two. In many companies, there can be a bias against processes at higher levels in the organization. Persons who manage other people often feel that processes are beneath them and should only be applied to underlings doing menial work. This attitude is often unwarranted, gets in the way of processes being embraced across a company, and really shows how little many people know about processes. In one very large dysfunctional service company with more than five hundred service locations, about half of the district managers were surveyed about job knowledge and work habits. These jobs were somewhat complicated and involved safety, sales, fleet maintenance, driver productivity, customer service, local HR, accounting, and local government affairs. The survey showed that only a handful of managers out of the 250-plus who were surveyed even attempted to do the whole job. Most people did what they knew and liked to do. Past salespeople did sales, past accountants looked at the books, and past maintenance people spent their time in the shop.

Everything else was on automatic pilot as these people faked their jobs every day. Almost all of the managers spent more than 40 percent of their time on the computer reading and answering emails and looking at *whatever*. It appears that most of these managers could benefit from a process that showed them how to plan and manage their day and cover all the bases. While different things develop for them in any given day that make each day somewhat different, their responsibilities over a week are very repeatable and do lend themselves to being controlled by a process. Before closing the door on any job holder who thinks he should be exempt from processes, it is best practice to go through the job responsibilities and look for repetition and a standard way of doing something. In general, most employees could use *more* processes in their work lives rather than *fewer*.

Processes are really good at defining a job or at least a big piece of the job. If processes are developed in the right way and take advantage of available best practices, they give the employee a set of instructions that will enable them to do the job better than if they just did it *on their own*, doing what they thought was best. Another benefit of processes is that the manager can be confident that everyone doing the same job is trying to do it in the same way. This not only provides better consistency but it improves quality provided the process is properly managed. Processes also serve as the foundation for employee training. It is practically impossible to train people to do a complicated activity well if there is no process that anchors the work and serves as a post-training reference for the employee. And performance management (we talked about that in the previous chapter) is really difficult for many positions if there is no process or collection of processes that provide a standard for how the work should be performed.

The detractors of processes always complain that processes stifle creativity and produce a bureaucratic morass that can drown a company. That can be true if leaders let it happen. We often see this in government programs where there is a six-inch-thick manual for everything and nothing of value ever seems to get done. But the problem here is not processes but the people who create them and administer them. Prudent leaders and managers understand that processes are a tool and not an entity unto themselves. Like any tool, processes cost money to build and more money to use and support. You build them where you have a specific need and can see a tangible benefit that justifies the effort. When a process is no longer useful, you eliminate it or improve it until it is useful again. A process should never be bigger than it needs to be and there should never be more processes than are necessary to get the work done right. Anything more is a waste, and that flies in the face of what processes are trying to accomplish.

There is usually a quality piece to a process that may involve metrics, targets, and oversight as well as feedback from a superior. The intent here is not only to show the employee how to perform the task but to have a mechanism that preferably gives the employee feedback on performance in real time. Good companies are always looking for better and more efficient ways to communicate with employees on how they are doing—and also empower them to do much of the performance management themselves. Processes are a good mechanism for doing just that. You can provide the process, train the employee on the process, give the employee manuals or other material that can serve as a reference guide, and then allow the employee to use the process to help them self-manage their performance with perhaps less oversight being needed from a superior. In the ideal situation, the employee would come to the superior for help when they have problems with an activity or the results they are getting. None of this is possible without the process serving as the framework for the work to be done, evaluating the results, and managing everything that transpires.

So to summarize, processes are basically a set of instructions on how to perform a certain activity. If properly constructed, a process should provide the best way of doing the activity. They need to be sufficiently detailed so that there is no confusion about what is expected, nor is there opportunity for corner cutting without violating the process. Everyone involved in the activity, from the employee doing the task to people that employee interfaces with and people supervising the work, should be covered in the process. Anything that is repeatable can benefit from a process.

Management-level people often think they should be exempt from processes. That is not always true. Processes help you stay focused on what needs to be done and how to do it, and most people can benefit from that. There is a natural link between training and performance management and processes. In fact, it is very difficult to train someone effectively and then manage their performance if there is no process in the picture on how to do the work. And last, good processes often have a quality piece involving targets and metrics that enable the employee to monitor their own performance and make adjustments. If properly developed and then managed, processes are a powerful tool for a company trying to get something done. Good companies do not operate without processes. So what is wrong with the bad companies?

Why Bad Companies Hate Them

As we have said, good companies just naturally gravitate toward processes. They know that this is the way to do things effectively and efficiently. The leadership, managers, and the rank and file have "grown up" with processes and see them as the necessary tools that they are. They are naturally comfortable with them and cannot imagine operating without them. You will not find many process haters in the good company. They do not get in, and if they do, they do not last long. These people end up at bad companies, where they conveniently find that most everyone else is of a similar mind. As we said in chapter 2, bad companies and their people do not have a successful legacy with processes. Previous experiences have been bad, resulting in failure, frustration, waste of resources, negative comments from superiors, even termination. In their collective minds, processes are something other people do for some strange reason and, amazingly, those people claim to get something good out of it. It is all too mysterious, plus it has never worked for them. They have gotten through life without them and see no reason to ever get near a process again. Every time the word *process* comes up, something bad happens.

Processes demand commitment, accountability, teamwork, and management of what everyone is trying to do. People in bad companies do not gravitate toward any of this. They want to be left alone to do things the way they feel like doing them on any particular day. Processes do not let you do that. Within a process, there is a standard way of doing things. If everyone does their part, processes work and give much better results. The problem is that in a bad company where everyone is *good*—and *better* is not valued—few people feel that processes are worth all the fuss. They are just too complicated and require too much discipline and work, and things like training people, which no one wants to do.

Sometimes in the bad company, you may find a manager who has basically built his own processes. Maybe this person was exposed to processes in a previous company or else did the necessary digging on his own to at least figure out what processes were all about and how to build and use them. Probably the resulting processes do not qualify as world-class, but they get the job done and they are a lot better than the random automatic-pilot approach used by everyone else. It would be good if this manager could share this work with colleagues and start a movement toward processes at the grassroots level. That is not going to happen. This manager knows he is swimming against the cultural current of the company and it would be suicide to stand out by doing something innovative. The peer pressure of mediocrity in the bad company is incredibly strong and makes sure nothing good gets created and disseminated down on the front lines. As we have previously pointed out, you can only move bad companies starting from the top.

For some strange reason, there may exist a process within the bad company. Perhaps an energetic high-level manager got it started or it was installed by a consultant. One thing is certain, the Terrorists and Rebels are going to zero in on that process immediately. If these resisters cannot totally kill it, they will make sure that it is not properly administered or managed. Either way, in a short time, the process will either cease to exist or at least cease to have any effect on the company. The resisters have never found a process they did not want to kill.

The key thing to remember when looking at bad companies wrestling with processes is that the culture of the company is anathema to the whole concept of processes. If a well-meaning CEO wants to install processes in a bad company, he must start by changing the culture of that company from the top down. Once companies change what they value and how they behave, then it may be possible to educate them about processes. You cannot do it the other way around. We will talk more about that in the next section.

How to Build a Process

It was pointed out in chapter 9 on key initiatives that you cannot fix a bad company by doing key initiatives. They all fail. You have to fix the company first and get it in good enough shape and then you can do a key initiative. It is the same with processes. The leadership of bad companies often thinks they can build some new processes and everything will be fine. That is totally backward—but typical thinking for a bad company. There is a reason why we talked about establishing the right leadership in chapter 6, assembling the right team in chapter 7, and then transforming the culture in chapter 8, before we got to building key initiatives in chapter 9. Those pieces have to be in place first so that the basic infrastructure is there to make a key initiative or, in this case, an important process feasible. To do anything less is foolishness and a waste of resources.

Once the basic infrastructure is in place and functioning properly, company leadership can start thinking about improving how the company operates. Chapter 9 covered the key initiatives, which are change efforts that affect the entire company and how it functions. They are massive efforts that touch virtually everyone and can change virtually everything, or at least it seems that way. For change efforts that are of such a large scale, the CEO and his team have to be front and center and highly visible all the way through development, installation, and integration. Company leadership has to capture the hearts and minds of the workforce from top to bottom, and that can only be done by a CEO-led coalition.

Processes are smaller in scale, more surgical and focused, and affect fewer people. Unless you are in the group where the process is to be used, you may not even be aware that a new process is in the hopper. While key initiatives address how the whole company does a particular thing, a process is concerned with how a specific group of people performs a particular function. So although key initiatives and processes do involve some similar steps, because of the scale and the focus, the process has to be developed and delivered in a different way to be successful. We will talk about that next.

With many processes, the CEO or another senior executive may be involved initially to set the expectation and show support, but then the focus needs to move down to where the work is being performed. Unlike key initiatives, processes address the work in just part of the company where those affected have similar expertise and priorities. These people understand the issues and what needs to be done better than most of the senior leadership. If a leader can identify the problem and frame what needs to be done, these people can handle many of the details. As an example, a senior executive in a large transportation company was charged with making major improvements to the fleet-maintenance program. The organization had more than four hundred operations, and each had its own shop. There were few centralized processes, and an initial survey of the shops showed that their quality varied from fairly good (B) to totally unacceptable (F). The overall program was judged not to be very effective (C- to D+). The senior executive knew the program needed standardization and processes, so he brought the maintenance managers together to get the ball rolling. He started the meeting out by summarizing where the program was and what needed to be improved and why. And then he stated that virtually all the knowledge and best practices that were necessary to turn the program around were contained in that room. That executive knew that processes must be built by the people closest to the work and that there were enough best practices in the room to provide a foundation for many of the maintenance processes that were needed. Importing a whole program from the outside and shoving it down their throats was not going to work. Many of these maintenance managers were very capable people and all of them understood the subject matter well. Not to get them involved would be insulting and would not engage them in the effort to turn the maintenance program around. He did not have to educate them or even get them interested. All he had to do was get them moving as a unit. This executive went on to build a coalition of the better managers and those with the best practices, and they in turn built the overall program. The maintenance team at corporate then helped install the processes out in the field and train managers and their staffs in the details. In retrospect, the executive felt he should have gotten the coalition of maintenance managers more involved in the installation and training. He believed that there would have been fewer problems with acceptance and that things would have moved more quickly and effectively. However, the critical message here is that although this project nearly had the scale of a key initiative, it was totally focused on maintenance. Getting managers at the front lines involved in building and driving the program was a major factor in the quality of the effort and its impact on how things were done in the maintenance shops.

This is where processes are unique. While so many of the transformation efforts discussed so far in this book involve developing the ideas and methods at the top and then engaging the rank and file to embrace them, with processes the creative forces are present at the front lines. The key is getting them interested and focused and functioning as a team. Using the preceding case history as an example, here is the step-by-step *process* for using the horsepower of the rank and file to build, install, and integrate a new process into a company that has the necessary infrastructure to support that effort.

1. Initially, someone must identify the need for a particular process. This can come from almost any level in the company. In better companies, people who actually do the work, or are supervising the work, often identify the need. In less proficient companies the initiation usually comes from management. The message must move up to a level

in the organization where the receiver has the authority to make the decision to do something.

2. The next step is to assign responsibility for the effort to a leader/manager who has sufficient background and knowledge in the work area, has the respect of the people who will be affected by the new process, and has demonstrated the ability to lead such an effort to successful completion. Many process projects succeed or fail depending on who is driving the effort. A new process may be warranted and many people may actually want it, but it does involve change and we have already talked at length about the difficulties most people have with change in the workplace. Even building and delivering processes that virtually everyone wants is hard work. Many leaders make the mistake of getting sloppy at this point and just pick who is available or who appears to have some interest in the project. While interest is important, remember to pick someone who can do the job and do it well. Anything less is putting the project at risk.

3. This leader/manager who we will call the Project Manager must now start by digging into the issue in order to accomplish two things. The first is to understand the issue up close (where the work is being done) and also to identify people at the work level who can serve as resources for development and implementation of whatever is needed in the way of process. This involves talking to a lot of people at several levels and listening to their perspectives on how large and complex the issues are and what is needed for a correction. This foundation-building is key. It provides the basis for what must be done. Plus, the optics are important here. If the Project Manager does not really understand the issues or has been lazy in his fact-finding, it will be difficult to engage those affected by the work.

4. The Project Manager must now broadly scope the project and build the team. Scoping the project means determining what needs to be done at least in general terms. Is a new process really needed or can some existing process be modified? Is there expertise required that is not in the company? Will other parts of the company be affected? How long will all of this take? The Project Manager must have at least a sufficient understanding of these issues so that he knows what to expect and can communicate that up the organization. Next, the Project Manager must build his coalition. This team should include a mix of corporate specialists who have specific duties in the effort and those managers/workers from the front lines who have best practices or have demonstrated the interest and ability to take the program forward. This will be a heterogeneous mix of people with different backgrounds, ideas, and biases. The only things they have in common is that they all work for the company and they are all closely involved in the work that the new process or modified old process will address. Picking people who can work together, resolve their differences, and stay focused on what is needed— and then produce—is vital. One disruptive individual can destroy the project, so the Project Manager must be vigilant in his selections and then move quickly if mistakes are made. The number of people on the team is basically a function of the complexity and magnitude of the work. One simple process can be covered by four to six people, while a project involving multiple processes or a very complex process could require twenty or more people working in a number of subgroups. It is important not to overcomplicate the effort but still to get a sufficiently broad input so that the quality is there.

5. Either before or at the beginning of the first meeting, the team must hear a message from an authority figure that communicates a sense of urgency and is also a call to action. We have covered this ground before when talking

about the CEO starting any transformation effort with a similar message. This is a smaller scale and a more focused group with narrower issues, but the need is the same. They must hear what is expected and why it is so important. The authority figure may not be the CEO, but he must be someone who everyone in the room knows and respects, and someone who understands the issues on the table. It must be clear what the marching orders are and that this person will be following up.

6. Next, the Project Manager must make sure the team is focused and is actually functioning as a team. This can be difficult and is often a stumbling block if not done correctly. He should give all the background on the project from the reasons for initiation to what he has learned during the due diligence phase. Procedures, ground rules, and expectations should be set and introductions should be made. Full disclosure is important here, so that everyone feels they are bona fide members of the team. The Project Manager should make it clear that he has ultimate responsibility for producing a quality product and that he is invested in this effort but that he has assembled the team in the room because he has confidence they are up to the task and he is depending on them. Independent of actual rank in the company, team members must feel that they are functioning as peers in the room and that all opinions have equal importance. The team members are there because they are important to the project. The Project Manager should lead the effort but with a light touch. Team members should be allowed to collectively develop their own team dynamic and the Project Manager should only exercise authority when significant problems develop or things really get off track. Even then, the intervention should be done with great care. The basic message is that everyone is equal and all opinions are valued. Things will begin to gel when team members start to have pride in what they are trying to accomplish.

7. The next part is a subset of step 6, but it is particularly critical and very tricky. Although the Project Manager is in charge and has responsibility for the final product, he has to build ownership and empower creativity within the group. That is the only way to get a quality product that can stick. This will not happen if the Project Manager acts like the smartest person in the room and micromanages the effort. The deft Project Manager should be managing what transpires in the room without appearing to be managing anything. To most people, he should be seen as just another team member. That involves paying close attention to what is being said, how it is being said, and what is not being said, and then very subtly acting on what is being observed. Perhaps some sidebar conversations are needed to modify perspectives, encourage someone to expand on or develop a promising idea, or draw out a more reticent participant. The Project Manager can offer up an idea or a direction of thought to move things onto firmer ground, but at no time should people feel that the Project Manager is making the decisions and they are just figureheads. The Project Manager needs to get the most out of each team member, as well as the whole collective team, and that takes observation and intelligent coaching. The desired outcome is every participant leaning in with elbows on the table, listening attentively and then making comments and offering up new ideas. If many of the team members have clammed up and are sitting back with their arms crossed over their chests, then you have lost them. A good Project Manager is watching for all of this. Another key component within this management effort is learning about and evaluating people. During the course of events, some people will impress and others will disappoint. The smart Project Manager will want to know the reasons for these observations before acting on them. Is the good performer genuine or just a practiced Terrorist intent on derailing events at the most opportune time? Was a mistake made by bringing in the underperformer—or are

there problems with the subject matter, the setting, or the mix of team members the person in question is working with? The Project Manager should be open to changing a few people or moving people around, but it should not be common practice and it must be done very carefully.

8. Once the team is assembled and brought up to speed on what is to be done, it is time to do some work. The Project Manager or some other team member can define the issues and frame expectations to get the ball rolling. This person can also serve as the moderator to make sure things flow smoothly and ideas as well as actual work product are properly documented. The key here is to guide the effort, but not hijack it. The team should have the freedom to pursue whatever "lands on the table." The final product must meet the standards of the Project Manager, but the team must also feel that they own it.

9. The next step is to put the process in final form so that it can be reviewed and ultimately used. There should be corporate members of the team or other people on the corporate staff who are capable process writers and can assemble the work product. However, if there are people from the field on the team who are interested in helping, it is a good idea to get them involved and maintain the field connection to the process. When the process has been written, it should be reviewed for content by the team members and then reviewed by someone *not on the team* for format and logic. The Project Manager should control this editing step and make sure the team is kept in the loop until the process is finally accepted. How changes are handled can be a potential stumbling block. The Project Manager has carefully managed the team members to get them to this point in the project, and that all can be ruined if the Project Manager gets autocratic about changes and does not take the time to hear all the opinions and deal evenly with the various points of view. The desired

outcome is to get a final product that all team members are proud of and feel they own, even if they are not totally behind all of the content.

10. Installation can be as simple as delivering the process to the users and being available to answer questions and then following up on progress. It can also be as complex as building an installation team of corporate and field members and using this group or several groups to train users on the basics of the process and then monitoring their progress in using the process. In either case, the key here is that users fully understand the process and can ultimately master all its details. It is possible, at this point, that a user may catch a mistake or logic bust that made it through the review stage. As part of the installation, the Project Manager must make sure the right people look at the issue and make the appropriate changes. This has to be done quickly and smoothly. Getting sloppy at this stage or letting egos get in the way can ruin what may have been a very promising project. Another issue during installation can be a field manager who is not in favor of the process and wants an exemption, wants to "water down" the process so that it is not effective, or just wants to undermine the whole effort. The Project Manager must be front and center on this and after "hearing out" the recalcitrant manager must either bring that manager in line directly or get more executive-level people involved. The important point here is that processes only work when everyone who needs them does them. When people get exemptions or become special cases, everything breaks down. Rebels and Terrorists know this and are always trying to be special cases.

The step-by-step procedures presented in the preceding pages are very detailed and there is a lot of emphasis placed on establishing an environment in the meeting room that engenders ownership and teamwork. The case history in the previous section required this because the team members were not initially comfortable with the process and there were trust issues between the corporate people and the rank-and-file workers. In the end, it worked out and a good product was delivered by the team. In more functional companies where processes are created and modified on a regular basis in virtually every area of the company, many of these steps are second nature and participants and managers hardly have to think about them. However, it is prudent to always be aware of the potential pitfalls and give enough detail to the effort to get what you want in the end, which is a workable process that justifies the effort.

Now we will talk about what it takes to get the most out of processes and keep them alive and functional.

How to Keep a Process Going

The integration part of the equation actually takes place well after installation. Companies make the mistake of thinking they are done once installation is complete. They disband the team, everyone goes back to doing their *real* jobs, and the process becomes an orphan. Sometimes that will work, but in most cases, that is a recipe for problems down the road. People start to modify the process on their own or decide not to use it. New users come in who have not been trained and they develop their own unique approach to how to use the process. Before long the only thing that is standard is the chaos created by this new situation as everyone does it their way. The way to avoid this is to appoint a process owner for life. This person is in charge of making sure the process is integrated into the company. That means the process is used in its accepted form by everyone in the company who should be using it and that all users are properly trained. People managing process users must routinely monitor their performance and deal with performance issues. If people want to modify the process, that request goes to the process owner, who can determine how best to address the issue. The process owner also monitors utilization through the managers, can arrange for additional training as needed, and can deal appropriately with compliance issues. In this way, the identity of the process stays intact, people across the company use the process as it was intended to be used, and any needed enforcement actions are identified quickly and handled just as quickly. When this happens, processes tend to stay alive and perform their intended function in an effective manner.

Great companies, and some good companies, take it a step further. They routinely look at the effectiveness of their processes and get the users directly involved in making needed adjustments. They use lean concepts to remove or modify steps that were originally well intended but have proven not to add any value. It is a lot like the "Work Out" approach at GE. Everything is put under the microscope and any step or form that does not add value is removed or changed. The new product is more efficient, more effective, easier to use, and cheaper to employ. The ultimate user can be confident that the process he uses is always state of the art and worthy of his time. If that is not the case, he knows how to work within the company to make the necessary changes.

The key thing to remember about processes is that they represent the better way of doing things. Good companies and their people know this. Good companies stay good because of them and bad companies cannot become good until they develop the attitudes and behaviors that appreciate and value processes. If the basic infrastructure is in place and processes are properly owned and managed during their creation, installation, and integration phases, then they will continue to add value until they are no longer needed. At that point, they will be replaced by newer and more powerful processes created by the same people who used the older processes and employed that knowledge base to make the needed advancements. The process of processes just goes on.

Chapter 14

Eliminating Time-Wasting Habits

Jim: So far in the book, we have discussed a wide range of attributes necessary to move an organization—the need for effective leadership, how to build a good team of managers and employees, how to promote a success culture—and we examined the three foundational pillars of good governance: safety, people, and performance management. In chapter 13 we covered key processes and how they help get the work done right the first time. However, none of this can be effective if people do not know how to get the most out of every workday. In this chapter, we identify some ways to help maximize productive time by reviewing a few of the most persistent time wasters that can hinder daily productivity.

There is little doubt that time is likely our most treasured asset. It is something to which we are all subordinate. Unlike things, time is a feature of life that can't be purchased, traded, manufactured, or replaced. Time is finite, and time is fleeting, and once gone, there is no rewind button to get it back. A wise person once said, "Sand is always running true through the hourglass. It doesn't stop for kings, and it doesn't stop for paupers. Be sure to use wisely the time you have been given."

As we said at the outset in the Introduction, if someone was stealing money or assets from an organization, you can bet that management would seek punishment for offenders and fix gaps that allowed it to happen. Yet those same managers may enable, and even participate in, wasteful routines that also negatively influence the bottom line. Why isn't there similar outrage when employees and staff are unable to maximize performance because they are occupied with inefficient and cumbersome routines that add little value in their jobs?

Whether deliberate or incidental, there is a tendency for time-wasting habits to slip into the social fabric of most organizations. Bad companies are really good at wasting time playing the office game as described in chapter 2. While managers may intellectually understand that time is a valued corporate asset, few connect the dots. And that can be very costly. A 2015 *Forbes* magazine article—"Wasting Time at Work: The Epidemic Continues," by Cheryl Conner—reported that, based upon data from Salary.com, the number of people who admit to wasting time at work every day "has reached a whopping 89%." The article relates that employees are spending longer periods than ever before wasting time. Their statistical breakdown:

- 31% waste roughly 30 minutes daily

- 31% waste roughly 1 hour daily

- 16% waste roughly 2 hours daily

- 6% waste roughly 3 hours daily

- 2% waste roughly 4 hours daily

- 2% waste 5 or more hours daily

Although there are many potential wasteful habits in companies, after years watching and being subject to such practices, we decided to focus in this chapter on some of the more significant time vampires that we have witnessed.

Meetings

In the last chapter we pointed out that in bad companies little attention is given to how to plan or conduct productive meetings.

This is supported by a number of credible studies that have shown that good meetings are uncommon in organizations today. In our experience, more time is wasted in poorly managed meetings than in any other daily business activity.

In contrast, in good companies, meetings can be a valuable management tool. They can be beneficial to update and convey critical information, align people around the mission, clarify roles and expectations, define deliverables, answer questions, address concerns, help streamline governance, and add structure to upcoming tasks. When properly planned and executed, meetings can have constructive impact on the bottom line. Effective meetings also serve to provide an enterprise perspective with all parties sitting around a table, which, in turn, can flatten contrived departmental silos. High-quality meetings can also help demonstrate in real time an appreciation for good foundational process discipline in keeping with the tone of our discussion in chapter 13. There is great benefit in requiring staff to organize thoughts and structure discussions. Good meetings can also provide an excellent teaching venue for the CEO and senior management to further fortify key leadership values, clarify expectations, and drive team alignment. They also enable senior management an opportunity to evaluate people—how orderly, organized, and efficient they are in presenting information and understanding strategy. To help define what we believe is effective meeting doctrine, there are several management principles we use to drive good meeting discipline:

1. Appreciate the optics—Meetings are not just for information exchange. Smart leaders use them as opportunities to anchor a culture of efficiency. To do that, sessions must exhibit the same order and centered practices that the CEO and top management expect in staff departments. This goes directly to our earlier point that what employees see has more fundamental impact on the culture than what managers say. For example, it is reasonable to believe that most managers openly assert expectations for crisp execution, disciplined processes, cost awareness, and attention to efficiency by workers in their daily jobs. This all

falls hollow, however, when those same managers preside over or participate in meetings that are unstructured, digressive, and without design. Staff will quickly pick up the contradiction.

2. <u>Understand the investment and the importance</u>— Remember that staff will put in a lot of effort in their preparation for a meeting. They may spend hours getting their presentations ready and script digested. It is a big deal to get face time and sit before the boss to talk about their projects. Give them the courtesy of clear purpose and plan in advance of the session. Too many times we have seen people prepare well for a meeting only to be disrespected by unorganized governance of the session.

3. <u>Invite only those necessary</u>—Only those with vested interest should be invited to meetings. Bad companies develop a "meeting culture" where everyone is invited and expected to attend—even those with limited or no association with the topic.

4. <u>Value the process</u>—There are several common attributes that we found relevant in conducting useful meetings. They include:

   - Define ahead of time the purpose, agenda, and desired outcomes of the session so that everyone is prepared and aligned on expectations. Then, stick to the agenda and don't wander.

   - Be on time and stay on schedule. It is as important to end a meeting efficiently as it is to conduct a meeting efficiently.

   - Consider a process facilitator to keep time and to guide meeting progression.

- Designate a keeper of record to document notes. Ensure you have a mechanism to follow up on action items. It is frustrating to come to a follow-up meeting and have to rehash material that was already discussed and resolved earlier.

- Develop rules of engagement to add structure and stay focused during the gathering. Pay attention to how the meeting will be structured and administered. In our experience, much effort is directed toward arriving at outcomes of a meeting, but little attention is paid to the process of the meeting itself. Without some common rules, discussions can digress into futility because people may go on tangents. Deliberations can grind down over a single point that may or may not be useful. Some rules of engagement that we have used to tee up good meetings include topics such as: 1) "storm inside—team outside" (i.e., everyone is free to air differences and opinions, but once a decision is reached, everyone will support the decision), 2) be on time, 3) no technology rule (keep personal electronic devices stowed), 4) no sidebar conversations, 5) be open to possibilities, 6) respect other views, and 7) share with complete candor. While these rules may seem obvious, we have found great value in keeping things moving to state expectations up front.

- Keep conversation focused on the topic and stay out of the weeds. It is frustrating to sit through a long discourse on someone else's micro-detail that doesn't widely apply. If a deep dive is needed on a specific topic, then take it offline or risk aggravating the rest of the team.

- Respect everyone's time by quickly quelling storytellers who dominate conversations with dialogue that takes the group on a tangent. Some people love to talk for the attention. Politely intervene and move on so the group remains engaged.

- Keep on topic and follow the agenda. Unless it is an open brainstorming session, you will want to limit free-lancing. Also, avoid the tendency for some to impress their peers with a dump of data versus presentation of information. We had a finance guy in the group in one organization who wanted to astonish the boss with his mastery over the numbers—even if the numbers had little to do with the topic at hand. It was painful to endure.

5. <u>Respect the logistics</u>—Choose a location that is suitable to group size and tenor. Sitting or standing in a room that is too small for the number of attendees blunts productive engagement. The physical discomfort can overcome attention spans and the significance of the material can be diminished. It is also important to ensure administrative supplies are available, and that needed technology is ready. There are few things more distracting to the flow and momentum of a meeting than to stop momentum to retrieve a flip chart or to manipulate a projector or electronic aide.

Sample Department Review Meeting Format

Any meeting has the potential to devolve into wasted time. We always tried to have a brief written agenda for every meeting—even one-on-ones with the boss. In our experience, however, business or department review meetings seemed especially prone to ramble and get sidetracked. So we developed a basic template that we used successfully with our teams. It is important to keep some flexibility in the process to be sure, but it is vital to keep structure intact to contain the discussion flow and maximize everyone's time. Our general review meeting guide included the following major categories in sequence of our discussion by department. This approach helped us keep everyone focused and consistent and took the guesswork out of staff preparation—in itself, a time saver.

Department/Group Review Meeting Sample Process Template

Top 3 Key Projects (force staff to prioritize focus on the most significant issues)
>> What is the project goal and purpose and how does the project help the department/company mission?

>> Where are we in the process, are we on schedule, and what is the plan going forward?

Metrics (key performance indicators that reveal performance)
>> Year over year to date—what is driving the numbers?

>> Current quarter vs. same quarter the year prior- what is driving the numbers?

>> Data analysis—what are the drivers of the numbers and how do they speak to us?

Significant Challenges
>> Any significant challenges/issues, and what is needed to overcome them?

>> What are the next steps, and what do you need from other groups?

Six-Month Look-Ahead
>> Present six-month look-ahead of items/projects coming down the line. Allows other departments to understand how they may be affected and/or may need to adjust priorities to maximize agency benefit.

Other Concerns, Comments, Questions

Meetings are a fact of life in most organizations. Effective, well-managed meetings are among the most visible hallmarks in a success culture. A meeting is an opportunity for management to demonstrate that process is important, that execution is important, and that managing time is important.

Email and Other Electronic Communications

Arguably one of the most significant communications enhancements in our time, email, when properly used, can dramatically improve productivity and streamline interaction. Until something better comes along, the reality is that email is here to stay. In most organizations, email has become the primary way of conducting daily business, communications, and routine interactions. There is no better means to quickly reach others, be it one person or a thousand or more people. One of the better features of email that adds value is the ability to transmit and share widely documents for information or review. The convenience of depositing a message or file into a colleague's inbox for later retrieval is a positive attribute. Despite time of day, the message will be waiting when convenient for the receiver. It is especially beneficial for firms with a workforce that is geographically or functionally dispersed.

However, like many good things in life, there can also be an unhappy side to email. In this case, despite the positives of e-communications, it can become a beast that overcomes even the most capable management teams. Instead of improving and speeding up exchanges, email, if not managed, can consume people and interfere with all aspects of life. Some people feel obligated to stay plugged in all the time, and some bosses think that email grants anytime access to employees, even if off the job. We have seen claims that four of ten bosses send emails to their people during non-office time. That number may be low in our experience. Some of those emails were information only, but many were sent expecting action and response—even during nighttime, on weekends, on holidays, and on vacations.

Some other common email dysfunctional routines that we have witnessed:

1. <u>Email[8] mania</u>—There are email maniacs in every organization. People who are addicted to emails and texts. They sit and stare at their incoming email page just waiting for the next message to pop in. And when there is a lull, they will reread messages. These people are addicted. They can't stop themselves from immediately checking the inbox when they hear the familiar *ding*, signifying message arrival. The addict brings about a certain degree of dysfunction by paralyzing productivity, attention to detail, and focus on the right things. Too many addicts in the group, and the organization goes into slow motion—or stops.

2. <u>Misplaced priorities</u>—To illustrate the depth of the electronic distraction problem, we sponsored a productivity assessment involving 1,500 or so frontline managers in a large operating company. The results were alarming: we found on average that managers were spending 40 percent of their time (or more) on electronic chores on computers (a good part of it responding to emails or data-reporting decrees from higher management) but less than 5 percent of their time on their people who were engaged in high-consequence and unforgiving operating tasks involving heavy moving equipment.

3. <u>Too much of a good thing</u>—Like any good thing, undisciplined email protocols can lead to overuse. People use email for everything, they copy the world, and they may get very long-winded. These are the people who use email to routinely talk to their colleague sitting in the same office or cubicle. They also include people on distribution who don't need to be included because they are not involved. Compounding the problem, some people respond to the originating sender by copying "all." Suddenly, the original

---

[8] We include text messaging in references to email.

email has babies and the babies have babies. The cumulative impact is email overload. Busy people get fed up and may not read the post. Some may even delete messages out of frustration without reading the detail due to the pure volume of material. That can lead to problems, since the significant information can be camouflaged by the irrelevant fluff and key assignments or communications may be missed. Time is wasted and people are out of the loop.

4. <u>Put up your electronic dukes—conflict and turmoil</u>—Email is cold and emotionless. It doesn't have benefit of gestures, body language, or voice inflections like face-to-face or voice-to-voice presentations. Hastily written emails can lead to turmoil and conflict. Words on the screen can be misread or misunderstood. In the heat of the moment, a tense email reaction leads to even more back and forth as players try to get a last point. All this consumes energy better served if directed at market issues or customer service.

5. <u>Wild goose chase</u>—Sometimes poor email context by the sender, or misreading by the reader, can lead to people heading off in the wrong direction. This is especially true if the unclear email is from a superior up the chain. People will scramble and go forward on what they think is expected—be it right or wrong. Effort is wasted and delivery of results is impaired or delayed.

6. <u>On the hook 24/7</u>—We mentioned this situation previously but believe it worth reemphasis here. Bosses who send urgent emails to workers around the clock risk creating burnout and resentment, not only from employees but also from employee spouses and families. Like the time a boss emailed a manager, while on a cruise vacation celebrating his anniversary, asking for a briefing paper on field staffing to justify positions. Needless to say, the vacation

was interrupted. The point was addressed by Jason Farman in his book *Delayed Response: The Art of Waiting from the Ancient to the Instant World*. In a December 11, 2018, Wharton School podcast posting, "Is *Waiting a Lost Art*" (knowledge@Wharton), Farman said, in part, "With the introduction of email, text messaging, with technologies that reach into our homes and our private lives, we feel really overwhelmed because we are expected to be available at all times...I think the result is this very overworked, burdened, busy population that is actually not as productive and not as innovative as it could be."

7. <u>The entrepreneurial personality</u>—This is the person who thinks the business is the most important thing in the world. He is delighted to work day and night and and may fail to appreciate that even their best people may not have the same sense of sacrifice and obsession for the work. Adding to the dilemma is the fact that entrepreneur-like thinkers are frequently full of new ideas and they are impulsive. It is easy to see how this can overwhelm staff with so-called urgent emails at any time of the day or night. The raining of endless emails (or texts or phone calls) intended to make the business better actually counters success. If employees never get time to unplug, and they feel like they're on call all the time, there will be some degree of resentment.

We recognize that sometimes a non–duty hour message is necessary for the mission and that senior-level managers or key employees may need to be available at any time for emergencies or critical, time-sensitive projects. But for routine matters, leave people alone. Great managers recognize that employees need to have a personal life in order to have the energy to engage and commit to the company while on the job. Here are a few techniques we have seen used to check the inclination to send emails after hours:

- <u>Set the example</u>—We have talked about the boss being on display. That goes here too. If you want your management team to limit email disruptions to staff off duty, then you need to do the same with them. Any policy you make needs to have you as the champion. The trickle-down impact will be immediate.

- <u>Differentiate</u>—Not everyone has to be available 24/7. Try to set some guidelines on what is urgent and what isn't urgent, and who needs to be involved. It is helpful to put yourself in the shoes of the person on the other end of the message. How would you receive it?

- <u>Be clear and concise</u>—Try not to send several emails on the same topic. Some bosses have a stream of consciousness on projects and send an email for every thought that comes. Instead, take a little time up front to formulate one email with all the facts before hitting send.

- <u>Get out the antacid</u>—Want to ruin the weekend for someone? Just send them a message riddle to the effect of "Call me as soon as you get in Monday morning about a problem" or "Be in my office as soon as you get back from vacation." You can be sure that this will cause distress and spoil time off for the recipients while they ponder what could be wrong.

- <u>Be sensitive to time of day</u>—Some bosses try to impress others by their work ethic. They send emails at 4 a.m. to show that they are on the job. Unless you are a twenty-four-hour operation, try waiting until a more reasonable time to hit send. It may save some murmuring and ridicule within your staff because they may read it as: "I am working around the clock, and you should too."

The importance of electronic communications in today's world is certain. And like other organizational processes, good companies recognize the importance of structured electronic practices. Bad companies don't bother. Perhaps most importantly, good managers set the example and they continually evaluate their teams to ensure electronic protocols benefit governance— not detract from it.

Other Common Time-wasters

As outlined in this chapter, poorly managed meetings and bad email customs and practices are among the most significant organizational time-wasters. But they are not the only ones. For example, here are a couple of other common habits that can deteriorate organizational effectiveness:

And the award goes to . . .—In every organization there are people who, if there isn't enough drama in life, create their own theater. Instead of collaborating and finding solutions, these people wear down the good people by constantly defaulting to negatives, rumors, and gossip. You can't have a conversation about anything work-related without the "Ain't it awful" discussion. These people waste an enormous amount of time while the good people grow frustrated.

The "wanderer"—In the early 1960s, the pop singer Dion sang about the wanderer. The song told the story of a young man who would "never settle down ....I move around, around, around." We have seen people like the wanderer waste time in places we have worked. These are the people who wander the halls looking for someone to visit. They are, as the sixteenth-century English idiom put it, the "hail fellows well met." They spend a good part of their day visiting with anyone who will accept them. Like a senior-level colleague we knew who considered it his mission to check in on other managers. You could hear him coming down the hall, and it meant you had better shut your door or you would find yourself engaged in a trivial, long-winded discourse about nothing productive. For busy people, this is highly distracting and annoying. In this case, since the wanderer was a senior-level person, subordinate mangers felt restrained from gracefully escaping the dialogue. As a result, work slowed and momentum was lost.

Reordering the Culture

We started this chapter by talking about the central impor-tance time plays in our lives. It is one of our most essential gifts, yet we seem to take for granted that time will always be there.

In our careers, we learned that time-wasting habits and cus-toms usually evolve. Rarely does someone set out to waste time in meetings or overuse email. Instead, the tendency to squander time seems to creep insidiously into organizational culture where it soon becomes the norm. People on the inside grow numb to the dysfunction and their definition of good is altered for the worse. The product of all this is diminished efficiency and output, higher costs, and reduced capacity. Failure of management to act to stem time-wasters does a disservice to shareholders and stakeholders.

How to do it Right

Limiting the negative impact of time-wasting habits is like any other organizational issue—it takes management awareness and a willingness to intervene. Good leaders watch for the natural ten-dency of people to slack off from time to time. While occasional brief diversions from the grind can be expected, when the digres-sions are chronic, become the norm, or are widespread, that's when problems arise. To help manage and control the diversions, we suggest forcing yourself to plan key must-do items for the day and the week and prioritize the list and conduct a truthful post-mortem for the day/week. Ask yourself if you got done what you planned and, if not, why not and what you need to do to avoid falling short going forward. Be open about your assessment and expect the same practice from your team members. It is surprising how a little attention to detail here can make a huge difference.

Conclusion

Making It Happen

"Good ideas are not adopted automatically. They must be driven into practice with courageous patience. Once implemented they can be easily overturned or subverted through apathy or lack of follow-up ....a continuous effort is required."

—Admiral Hyman G. Rickover

Jim: By this point in the book, we hope it is clear that orchestrating lasting change and moving from a culture of failure to a culture of success requires both art (the leadership qualities that engage the team and fuel the change) and science (the processes, systems, and metrics that define the architecture of the change). Our goal in writing this book was to present a balanced personal appraisal of both sides of the equation—the art and the science of leading change and, most relevantly, how the two realms connect. We have outlined in these pages how to recognize dysfunctional creep, how to confront it, how to curb the negative inertia it brings, and why it is important to do so—to enable your organization to maximize value to all stakeholders.

As we stated early on, fixing bad companies involves leadership resolve and must be generated from the top. In chapter 6 we discussed at length that the CEO and his team must be ever mindful that they cannot take their foot off the gas and they must keep communicating key messages to the rank and file. We said that the CEO and his team must own the transformation and the culture. We said that driving positive transformation is a never-ending process that has a beginning but does not have an ending. A winning culture must be cultivated, fed, and cared for by the leadership team.

Fixing bad companies also requires several component parts that must all be included if a transformation is to be successful: building the guiding coalition, addressing negative culture, telling the truth, establishing the right key initiatives, transforming safety, defining the role of HR in upgrading the workforce, creating a performance-management capability, and building and supporting key processes. These are necessary components to successfully take the company from "bad" to "good," which translates into higher margins, more efficiency and capacity to grow, and much higher employee allegiance and customer loyalty. There are no shortcuts. Each of these components is vital and cannot be eliminated or done partially. What we have learned is that bad companies do not like change and the people in those bad companies are very good at derailing efforts to change anything in the company.

We pointed out that the best companies have engaged leadership, alignment in vision, clear accountability, good processes, effective metrics, healthy team interaction, an overriding passion to improve, and, as discussed in chapter 3, commitment to telling the truth. All of it must be driven from the top as the CEO and senior leadership team openly take the role as guardians of the culture. Great leaders understand that aligned, engaged employees create a highly competitive advantage in today's marketplace. Customers are better served, resources and costs are better managed, and differential effort is given at all levels.

In this final chapter, we recap some of the underlying themes and primary leadership qualities discussed in preceding chapters that, in our experience, work best to establish and anchor a winning culture in complex and sometimes highly reluctant organizations.

To summarize, in our view, qualities of great leadership are defined by people who:

Display character—Character shows through no matter ability, talent, or circumstances—and that is crucially important. People want to be proud of their boss. Humility, selflessness, loyalty, determination, positive energy. All are qualities of great leaders.

Align the team—In chapter 6 we emphasized that it is the CEO and senior team who must ensure that everyone in the company understands, buys in, and aligns with the values and the vision. Good companies consistently question what they do and why they do it. Candid conversation is encouraged, and people feel free to challenge each other to ensure an external focus so work activities translate into value-add for the customer.

Own performance and results—The leadership team is responsible for creating a culture of engagement in an organization. It starts with the senior team and migrates to the front line with consistent, cogent, and clear direction so that everyone is on the same page of the playbook. Leaders display character, competence, courage, and empathy—a *follow me* style, leading by example, empowering people, and teaching accountability.

Ensure mid-level management buy-in—Obtaining and sustaining support, alignment, and buy-in from frontline/middle management is essential for long-term success. As the interface between strategy and execution, mid-level people keep the organizational mission on track every day. They are closest to employees, customers, regulators, suppliers, and the public at large. They channel informational flow up and down the organization. They have enormous power in the enterprise. You want your mid-level management team on board with you.

Maximize communications, messaging, and branding—People in an organization have a desire to be part of something special. To be part of a group with special rights of passage where people can't just come and go at will. You must earn your right to be part of the team. This helps create a mystique, and people take pride in belonging. Ownership and pride bring improved performance and higher profits.

Deal with internal skepticism—There is always some degree of skepticism in the ranks when there is change or new direction. People inside the organization will more readily buy in and support a plan or an idea if they see a direct-line benefit to them.

Maintain a sense of urgency—Incite emotion around the vision. There must be a compelling reason presented to people to buy into the vision. They must understand why you are doing this and why it benefits each one. A leadership display of positive energy with a sense of urgency and obsession with improvement will invigorate a team to action. People are attracted to dynamic leaders. Most people want to see that their efforts are producing a positive impact. Smart managers mix a sense of urgency and aura of confidence to excite their staffs to action. The shareholders and customers win big.

Measure and create a heaven and hell—Good people want to be measured. Bad people don't. Good people take pride in producing results and demonstrating them with metrics. Bad people hide behind organizational "trees" and don't like measures that eliminate the hiding place. Great organizations measure performance, process, and execution. They create a "a heaven for those who align and engage, and a "hell" for those who don't.

Execute and promote accountability—Anchoring the culture requires precision execution to stabilize governance and establish defined ways of executing in the culture. As a military fighter pilot, Jim learned early that the one who executes with precision lives to fly another day. Military fighter pilots have certain methodologies that help them execute with accuracy. Like fighter aviation, companies that best execute in the competitive marketplace

271

are those that prosper. That consistency promotes a culture where things are done the right way based upon the values.

Guard against management hubris—Organizational performance can decline if leaders allow hubris to take root (i.e., self-centered and arrogant management, underestimating the competition, failure to recognize change, failure to recognize risk by redefining good to suit personal agendas, etc.). In an October 9, 2018, *Wall Street Journal* article, Sue Shellenbarger wrote that research has demonstrated that humility is a core quality of leaders who inspire close teamwork, rapid learning, and high performance in their teams. She goes on to say that companies with humble chief executives are more likely than others to have upper-management teams that work smoothly together and share decision-making. The bottom line: stay humble—you can always do better.

Get the people right—In several chapters we articulated the criticality of getting the right people in the right roles. Jack Welch was credited with saying, "Nothing matters more in winning than getting the right people on the field. All the clever strategies and advanced technologies in the world are nowhere near as effective without great people to put them to work." Great leaders understand the importance of instituting the right team architecture. Regularly evaluate your people, their attitudes, capabilities, and suitability for the task ahead. You need the best people in the right place to maximize operational and financial performance. Force-ranking people is a healthy exercise. The process alone provides many positives for an organization. For one, ranking people ensures that you put the right people in the right seats. It also forces fence-sitters to get engaged or risk elimination, as well as helps you identify possible future management stars.

Ensure that what is said is what is done—People believe what they see, not necessarily what they are told. For example, the CEO in a major transportation company said openly that safety was a core value and compliance with company standards was essential. One day, an area manager was observed by employees leaving the operating site with his cell phone in his hand, up to his ear, engaged in a conversation—in violation of company policy on use of cell phones while driving. To those who witnessed or later were made aware of the incident, the real message was clear—safety and compliance with company policy were not all that important. The CEO's words were instantly rendered empty.

Consistently drive values—There are many examples where we see leaders conceding their values and jeopardizing team loyalty, especially if it involves a star player or a highly tenured employee. Allowing one individual to get away with putting himself above the good of the team is a death sentence for leadership credibility. Letting standards slip sends a strong and clear message to everyone else that values are to be observed only when it benefits the circumstance. This guarantees loss of team essence and demotivation of the good people who follow the rules and put team first.

Insist on high standards—We believe that leadership insistence on high standards is a good way to motivate employees. Low standards lead to low performance and low commitment. High standards can energize the team and create a sense of pride in belonging.

Be obsessed with improvement—To make a point, keep in mind that a walrus isn't born fat and ugly – it gets that way over time. It is the same for businesses. Organizational cholesterol develops and invades healthy processes. Unless there is a leadership obsession with improvement, a company risks the gradual onset of complacency and loss of edge.

Clarify the vision—Someone once said that people first buy into the leader before they buy into the vision. This may be a good time to reiterate what we mean by the term "vision" as it relates to organizational life: the vision is what the group aspires to be .... the ultimate benchmark of success. In order to gain alignment and ensure everyone is pulling in the same direction, people at all levels of the organization need to understand the vision and their role in helping the company achieve it. The vision must be clear, concise, and easily understood by everyone in the organization.

<u>Manage conflicting customs and lore</u>—Too many organizations we have seen may have formal processes well outlined. But the informal processes—the way things are done—do not align. Legend, lore, and custom slowly consume formal practices. Good companies ensure that there is total alignment of all processes—those prescribed as well as those non-prescribed.

<u>Recognize and reward</u>—Anchoring the right culture requires that formal recognition and reward programs relate and ally with the expected code of conduct. For example, in a large service company an attempt to institute a culture of customer engagement was commanded from the C-suite. Programs were enhanced, metrics were developed, and internal communications were focused on the importance of the initiative. However, the recognition and incentive plans were not adjusted to reflect the new push. As a result, management was rewarded under the previous criteria that didn't include any of the new priorities around customer-engagement activities, such as proactive contact when service was disrupted, reduction in the number of calls it took to resolve an issue, etc. The lack of alignment between the new focus and management incentive targets presented mixed messages and marginalized the impact and success of the new program.

<u>Appreciate that it is a journey not a destination</u>—A culture of engagement is not a destination. It is not something that you achieve then go on to the next challenge. Building and sustaining a culture of positive engagement is a journey that never ends. As we discussed, the real benefit is that the journey itself is as important as the outcome. The process of going through it will drive success. Striving to get better every day provides energy that powers improvement.

<u>Don't assume that things are fine</u>—Business life is dynamic and fluid. There will be change, and nothing stays the same. Whether

the inevitable change is good or bad is up to the leader. You can't get complacent. Every day counts in nourishing solid team character. Good leaders constantly assess and ask themselves and their teams, "Where are our gaps?"

Stay the course, even when it hurts—Consistency in values and message will define the sustainability of your culture. To your people you are the owner of company culture. That is why it is important to stay the course and focus on taking care of issues that challenge the culture. Being consistent, no matter the challenge at hand, goes a long way to anchor the right culture with employees and stakeholders. In his November 2015 article "What Amazing Bosses Do Differently" (*Harvard Business Review*), Sydney Finkelstein wrote, "Who could be happy with a boss who does one thing one day and another thing the next? It's hard to feel motivated when the bar is always shifting in unpredictable ways and you never know what to expect or how to get ahead. So be consistent in your management style, vision, expectations, feedback and openness to new ideas. If change becomes necessary, acknowledge it openly and quickly."

Never take good employees for granted and never assume that they will always be there—Never compromise on baseline values, because team cohesion may be at stake if there is inconsistent messaging or actions. Good employees need TLC too, and they need to see their leaders confront nonperformers. Don't worry about motivating the poor performers. Worry instead about not demotivating the good ones who make it work every day.

Eliminate the Rebels and the Terrorists—In an April 10, 2018, *Harvard Business Review* article, Abby Churnow-Chavez pointed out that research revealed a 70 percent variance between lowest-performing teams and highest-performing teams correlated to the quality of team relationships. She wrote, "One toxic team member is all it takes to destroy a high performing team... toxic team members are destructive because they create unnecessary drama, erode the team brand, undermine the values of the leader and the company, and degrade team culture." As detailed in chapters 6 and 7, and elsewhere in the book, our experience is that unless the Rebels and Terrorists are removed, they will literally kill any successful transformation.

Listen to the organization—Recognize that the organization will talk to you, so listen and watch for signals that reveal organizational climate. Reality is your friend. Pay attention and be quick to react if you sense that the culture is getting lazy.

Watch for mission drift—When senior managers get lazy, there is loss of focus on the mission. Pete Steinke calls this *mission drift* in his book *A Door Set Open: Grounding Change in Mission and Hope*. He points out that mission drift is when people lose sight of the reason they come to work every day. "They fool themselves that they are vital or viable simply because they endure, preoccupying themselves with nonessentials." Along the way, the mission of the company and serving customers fades.

Establish pride in the craft—One of the great blessings about being in a leadership role is the ability to provide people a higher calling in their work. Elevate the way your people view themselves and the company role in making the world a better place.

Celebrate the "skid marks"—Great leaders reward the right behaviors. It is critical to recognize and celebrate the people and behaviors that affirm the desired culture. In anchoring a positive culture, hire and promote the believers. Celebrate widely even small victories that affirm the behaviors you are trying to moor into the social fabric. Recognize and promote those who live by the values and support the mission in the right ways. Be quick to capitalize on good ideas that come from those who buy in and work at it. Make it socially unacceptable to be an outlier in the new culture. Typically, we have found that about 60 percent of the people in a transformational effort will jump on board with you quickly, about 20 percent will wait to see what develops and can go either way, and about 20 percent just won't move with you. When you celebrate the "skid marks" in the new culture, you provide the tipping point that will secure most of the 20 percent who are on the fence. Manage to the majority, and don't waste your time on the 20 percent who don't want to get it. Adjust quickly in that segment to avoid distractions and the negative inertia they bring to the ranks. Remain vigilant to those toxic people and don't let them get in the way. Letting dysfunction stew in hopes it will go away does not work. Good leaders keep the pressure on. Be obsessed with improvement and measure the climate regularly. It takes energy and will to drive a positive culture through the ranks.

Recognize that people value most that which they cannot buy—Theodore Roosevelt was credited with saying, "Nobody cares how much you know until they know how much you care." More recently, Clarence Francis, former chairman of the board at General Foods Corporation, was credited with saying, "You can buy a person's time; you can buy their physical presence at a given place; you can even buy a measured number of their skilled muscular motions per hour. But you cannot buy enthusiasm, you cannot buy loyalty, you cannot buy the devotion of hearts, minds, or souls. You must earn these." More than just common sense, there is enough research out there to state with some certainty that general job satisfaction isn't always about the money. While important to pay the bills, money can't buy the things people find of most value.

Don't forget who packs your parachute—Going hand in hand with the immediately preceding paragraph, I often tell the following story in my keynote presentations about a life lesson that I learned early in my professional life while serving as a United States Air Force pilot. It was an encounter that underscores the importance of being a humble leader.

The son of parents from long-established working-class families, I was blessed to be raised in a home with a strong blue-collar ethic. I developed a special understanding of those who toil without complaint in the shadows with little glory or recognition. I was the first on either side of the family to graduate from the university. During my time at school, I developed an interest in flying fast airplanes. After four years of military leadership training, academic testing, and physical exams, I was selected, upon college graduation, to attend USAF pilot training after a very rigid and extensive selection process with only a handful selected out of every one hundred applicants.

Once commissioned and on active duty, it was on to a base in Texas for flight school. The competition was stiff. Everyone there had gone through similar screening just to get the appointment. The curriculum was pressure-packed and aggressive. Sweat is a great solvent, and program washout rates were high as the training deepened. I was on a mission. I applied myself and did well in all phases of training, graduating high in the class. Based upon academics, flying scores, and physical fitness, I was able to secure a highly coveted assignment flying the top-of-the line jet fighter of the era—the supersonic Mach 2+ F4 Phantom II.

In the Phantom squadron, every pilot was at the top of the pyramid. It made for very high comradeship and aloofness. I was at the top of the food chain ....or so I thought. Unfortunately, somewhere in the journey from leaving home to that assignment flying an F4, I lost my way. I was set up for a very hard and humbling lesson. One day, walking down the street on base, I saw a sergeant on the other side of the street. I didn't acknowledge him . . . after all, I was a fighter pilot. I didn't think I had anything to say to anyone who didn't fly a "fast mover." The sergeant crossed the street, gave me a salute, and said, "Sir .... I am Sgt. Winslow; I am in your squadron." I said, "Sgt. Winslow, I don't recognize you. What do you do in the squadron?" He said, "Sir, I am responsible for packing your parachute."

He packed my parachute! If something went wrong in the jet and I needed to jettison my airplane, my life depended upon Sgt. Winslow doing his job well. Yet, so consumed with my own ego, never had I taken time to notice or appreciate those, like Sgt. Winslow, who were critical to my success. How could I forget that I was a member of a team and that I was standing on the shoulders of people who labored every day in the trenches so I could succeed? Since then, I have tried never to forget that people pack my parachute every day in so many ways. Leadership is a higher calling, and we can never forget the power of the people around us.

In summary, our experience is that the lack of leadership conviction destroys any chance of instituting a culture of positive energy in a company. And a lack of energy produces poor safety, lost productivity, and financial underperformance. Just as significant is the potential collateral damage suffered through loss of social reputation and corporate credibility with customer, community, and other key stakeholder groups. Loss of reputation brings loss of trust. This can be a going-out-of-business strategy. A Watson Wyatt research finding that was quoted in a 2015 Edelman Corporate report on reputation management revealed a stunning fact: companies with high employee trust levels outperform those with low trust levels by 186 percent (based on three-year total shareholder return). This is a remarkable endorsement why leadership conviction and treating people in a just manner helps build trust and correlates to big financial benefits.

Throughout the book, we have emphasized that fixing organizational dysfunction and building a success culture is not a destination. It is not something that you achieve then go on to the next challenge. As a leader, once you think your company is *"there,"* you most certainly are setting yourself up for dysfunctional creep that can make your culture sick. Building and sustaining a success culture requires nurturing with continuous leadership focus. It is a journey worth taking as a differentiator that helps ensure maximum effectiveness and financial performance in any company, and we hope that the tactics and strategies that we have outlined in this book, and summarized in this chapter, will help leaders and managers anchor positive, productive, and healthy organizational cultures and take their companies to the next level.

Notes

Introduction

1. James C. Collins, *How the Mighty Fall and Why Some Companies Never Give In* (New York: Harper Collins, 2009)

2. William Bruce Cameron, *Informal Sociology: A Casual Introduction to Sociological Thinking* (New York: Random House, 1963)

3. National Association of Corporate Directors "NACD Audit Committee and Compensation Committee Chair Advisory Council: Nonfinancial Metrics, Strategy, and Culture" (Issue Paper, February 15, 2018)

Chapter 1

1. Jim Collins, *Good to Great* (New York: Harper Collins, 2001)

2. Larry Bossidy and Ram Charan, *Execution: The Discipline of Getting Things Done* (New York: Crown Business, 2002)

3. Jeff Weiner interviewed by Gayle King on "CBS This Morning" October 11, 2017

Chapter 2

1. Kati Najipoor-Schutte and Dick Patton, "Survey: 68% of CEOs Admit They Weren't Fully Prepared for the Job" (Harvard Business Review, July 20, 2018)

Chapter 4

1. John P. Kotter, *Leading Change* (Cambridge: Harvard Business School Press, 1988)

2. Larry Bossidy and Ram Charan, *Execution: The Discipline of Getting Things Done* (New York: Crown Business, 2002)

Chapter 5

1. Pierre Beaudoin, "Flying people, not planes: The CEO of Bombardier on building a world-class culture" (McKinsey Quarterly, March 2011)

2. Alicia Bassuk, "How to Deal with An Office Soapboxer" (Harvard Business Review August 30, 2016)

3. Brian L. Fielkow, *Driving to Perfection: Achieving Business Excellence by Creating a Vibrant Culture* (Minneapolis: Two Harbors Press, 2013)

4. Len Fisher, *Rock, Paper, Scissors: Game Theory in Everyday Life* (New York: Basic Books, 2008)

5. Right Management Study "The Power of Recognition" (Towers Watson, 2009)

Chapter 7

1. Amy Elisa Jackson, "Why Southwest Says Soft Skills Reign Supreme" (Glassdoor.com, December 4, 2018)

2. John P. Kotter, *Leading Change* (Cambridge: Harvard Business School Press, 1988)

3. John P. Kotter, "What Leaders Really do" (Harvard Business Review, 2001)

4. Christine Porath, "Isolate Toxic Employees to Reduce Their Negative Effects" (Harvard Business Review, November 14, 2016)

5. Rod Collins, "How the Best Company to Work for Works" (Management Issues Magazine, November 16, 2016)

Chapter 8
1. Jeffrey Pfeiffer, (Harvard Business Review, 2011)

2. Jack Welch, (Brainy Quote https://www.brainyquote.com/authors/jack_welch).

3. Patriot Rail Company Values Statement, 2017.

4. Brimstone Consulting Group, "Accelerating Results Through Alignment" (2017)

5. James O'Toole and Warren Bennis, "A Culture of Candor" (Harvard Business Review, 2009)

6. Steven M. R. Covey, "How the Best Leaders Build Trust" (Leadership Now/M2 Communications, 2009)

7. Bob Marcus, "Creating Candor in your Organization" (Nvolv, 2018)

8. Loren Steffy, "Up in the Air" (Texas Monthly, November 2015).

Chapter 9

1. Robert Slater, *The GE Way Fieldbook* (New York: McGraw-Hill, 2000)

2. John P. Kotter, *Leading Change* (Cambridge: Harvard Business School Press, 1988)

3. Gordon Bethune, *From Worst to First* (New York: John Wiley & Sons, Inc., 1998)

4. Larry Bossidy and Ram Charan, *Execution: The Discipline of Getting Things Done* (New York: Crown Business, 2002)

Chapter 10

1. Hon. Robert L. Sumwalt, presentation to Patriot Rail Company senior leadership workshop in November 2013.

2. Tony Kern, *Darker Shades of Blue, The Rogue Pilot* (New York: McGraw Hill 1999)

3. ICAO SMS: https://rise.articulate.com/share/ v5Sm_ODJQvKI51ZQb6HJmBy7bOrhQfTE#/

4. NTSB Aircraft Accident Report (AAR-14/01), "Descent Below Visual Glide Path and Impact with Seawall: Asiana Airlines Flight 214; Boeing 777-200ER, HL7742."

5. James Reason, *Organizational Accidents Revisited* (and other multiple references) (Boca Raton, London, New York: CRC Press, 2016)

6. Diane Vaughan, *The Challenger Launch Decision* (Chicago: University of Chicago Press 1996)

7. Mike Mullane, NASA (Retired) presentation to "Leading People Safely" conference, Houston, Texas, September 22, 2016

8. Hon. Robert L. Sumwalt, presentation to Air Line Pilots Association 2007 International Safety Forum, Washington DC

Chapter 11
1. Larry Bossidy and Ram Charan, *Execution: The Discipline of Getting Things Done* (New York: Crown Business, 2002)

Chapter 12
1. Larry Bossidy and Ram Charan, *Execution: The Discipline of Getting Things Done* (New York: Crown Business, 2002)

Chapter 13
1. Robert Slater, *The GE Way Fieldbook* (New York: McGraw-Hill, 2000)

Chapter 14
1. Cheryl Connor, "Wasting Time at Work: The Epidemic Continues" (Forbes Magazine, July 31,2015)

2. Jason Farman, *Delayed Response: The Art of Waiting from the Ancient to the Instant World* (New Haven: Yale University Press, 2018)

Conclusion
1. Sue Shellenbarger, "The Best Bosses are Humble Bosses" (Wall Street Journal, October 9, 2018)

2. Sydney Finkelstein, "What Amazing Bosses Do Differently," (Harvard Business Review, November 27, 2015)

3. Abby Churnow-Chavez, "Ways to Deal with a Toxic Coworker" (Harvard Business Review, April 10, 2018)

4. Peter L. Steinke, *A Door Set Open: Grounding Change in Mission and Hope* (Herndon, Virginia: Alban Institute 2010)

5. Edelman, "Activating Communications Marketing— Promoting, Protecting & Evolving Reputation"(Corporate Reputation Presentation, 2015)

About the Authors

Chuck and Jim have been close colleagues and friends for almost two decades. For seven years, they worked side by side as senior executives for a Fortune 200 company, jointly leading efforts to transform performance in safety, operations, and productivity. For example, they instituted and led programs that reduced worker casualties by 75 percent and workers' comp costs by more than 50 percent in just five years; instituted controls and protocols that saved more than $400 million in procurement costs on a $5 billion annual spend; and implemented a metric-driven process that improved productivity by more than 2 percent—bringing millions to the bottom line in both direct and indirect cost reductions. Since then they have worked together on a number of projects, bringing transformational change to underperforming organizations in both the private and public sectors. Today, Chuck and Jim continue to collaborate and team together in leadership consulting, keynote speaking, and coaching engagements in high-consequence industries.

About Charles E. 'Chuck' Williams

Chuck's career can best be described as a transition from engineering to the waste-management business and then to business consulting in the transportation industry. Chuck has three degrees in civil engineering from the Massachusetts Institute of Technology and began his career in engineering consulting. He was co-owner and principal of a Houston-based consulting engineering firm for fifteen years. Typical projects included high-rise commercial office buildings, port infrastructure, utility tunnels, pumping stations, industrial plants, and hazardous-waste cleanups. During his tenure, the firm grew to a 130-person organization with work in a number of states and on the international front before being sold to Raytheon. Chuck received numerous awards from industry groups for technical achievement and a number of his methods are still being used on major engineering projects. Chuck entered into the waste-management industry as a member of the startup team for an IPO. That original entity grew to more than forty times its original size before being sold. Chuck developed the technical department from scratch and was directly involved in the evaluation and integration of all acquisitions. Over the next three years, Chuck participated in four billion-dollar-plus mergers/acquisitions where he integrated various technical and operating groups, created new technical capability, and ultimately managed a technical team of more than five hundred specialists. He was presented with two achievement awards by the two national waste industry groups for technical innovations. In 2000 he became the Senior Vice President for Operations for Waste Management, a Fortune 200 corporation. During his seven-year tenure, safety incidents were reduced by more than 75 percent, more than $400 million in procurement savings were realized, and fleet maintenance costs for a 25,000-truck fleet were reduced by more than 20 percent. In addition, his efforts improved route productivity with a formalized methodology; and he created the industry's first operations scorecard. Chuck also took over management of the landfill gas to energy business and his team dramatically improved and expanded the business. Since that time, Chuck has worked as a management consultant, helping to turn around struggling entities in the rail and port industries. He has conducted executive team evaluations, made improvements in project management and productivity, coached senior executives, and built numerous processes for various applications. He learned many of the principles advocated in the book during his time in the waste industry and has been able to apply them with success to the turnarounds he has encountered in his consulting business. Chuck also has an executive degree from the Wharton School of Management.

Contact Chuck: charleswilliams@stx.rr.com // https://www. linkedin.com/in/chuck-williams-69789010/

About James T. 'Jim' Schultz

Jim is a well-established senior executive with more than forty-five years of hands-on experience leading transformational change in complex organizations. He developed early a passion for excellence while serving as a supersonic F4 Phantom II jet fighter pilot in the United States Air Force (USAF), where he acquired his "Don't forget who packs your parachute" approach to leadership. Today, Jim leads his own consulting/coaching practice with clients in high-consequence operations in the rail, trucking, aviation, power generation, maritime, and financial sectors. Prior to retirement, Jim held  various senior-level posts in both the private and public sectors, including Executive Vice President and Chief Administrative Officer at Patriot Rail Company; Senior Vice President Employee and Customer Engagement, and, Vice President Health and Safety at trucking giant Waste Management Inc. (WM); and, Vice President and Chief Safety Officer at the Class 1 railroad CSX, where he was termed "culture change guru" by *Progressive Railroading* magazine for leading a cultural reinvention after decades of adversarial labor union/company management interactions. Jim also offers a distinctly unique leadership perspective having been appointed to the Federal Senior Executive Service (SES) as the top career safety official for the Federal Railroad Administration (FRA), where he managed the nation's federal rail safety oversight and enforcement activities through a cadre of more than five hundred federal rail safety inspectors. During his tenure at WM, Jim's "Mission to Zero" (M2Z) and "Life Changer" programs, in five years, produced a marked improvement in foundational safety culture in this international 50,000-employee company, resulting in a dramatic 75 percent reduction in worker injuries and a more than 50 percent reduction in the corporation's annual workers' compensation cost. Jim and his programs have been featured in print media, including the *Wall Street Journal*, the *Washington Post*, *Traffic World*, the *Journal of Commerce*, and other public and industry publications. He is highlighted in the leadership book *Red Zone Management* (2001 Holland). He served as a rail safety spokesman on screen for TV news media outlets and as an expert consultant for ABC's *20/20*. He is coauthor with Brian Fielkow of *Leading People Safely: How to Win on the Business Battlefield* (North Loop Books, 2016); and he has authored articles and commentaries on rail safety, leadership, and culture, including several reports to Congress. Jim was a faculty lecturer in the European Divisions of the University of Maryland, and Central Texas College, while stationed in the USAF overseas. He received the civilian Silver Star for Bravery from the American Federation of Police for action while in the USAF. Active in the community, Jim was formerly a member of the Executive Committee, Chairman of the Personnel Committee, and a member of the Board of Directors for Jacksonville Urban League. Jim holds an FAA commercial pilot license certification with multi-engine jet and instrument ratings. He earned in residence a master's degree from Webster University, a bachelor's degree from Arizona State University, and completed postgraduate and executive leadership programs in residence at the Harvard Business School, the University of Southern California, and Northwestern University.

Contact Jim: jim@jimschultzgroup.com //http://www.linkedin.com/in/jimschultzgroup/

Additional Publications by James T. Schultz

L *eading People Safely: How to Win on the Business Battlefield*
       Schultz, James T. and Fielkow, Brian L. (Minneapolis: North Loop Books, 2016)

In *Leading People Safety*, the thesis is simple: safety is the decisive component in the social DNA of best-in-class companies. But safety is often marginalized and relegated to dense handbooks that are ignored by employees. *Leading People Safely* instead offers a straightforward how-to guide for maximizing organizational performance through safety leadership. The book demonstrates why safety must be a core value engrained into the rhythms of every task and shows how to bring people and process together in full alignment to provide a definitive competitive advantage. *Leading People Safely* is packed with succinct, savvy know-how for implementing a culture of safety, punctuated with easy-to-skim lists and textboxes. A 2016 Kirkus review stated, in part, "Schultz and Fielkow use stories from the authors' own experiences help to enliven the text. A fighter pilot turned corporate executive, Schultz has wide experience successfully leading transformational change in both public and private sector organizations based upon his trademark philosophy: 'Don't forget who packs your parachute.' Fielkow owns a transportation and logistics company, and the authors provide a lot of examples of how he implemented ideas outlined in the book. It is a valuable read for executives and business owners looking to establish a healthy company culture with a safe work environment."

Highly popular, *Leading People Safely* was named "Top New Release" on Amazon in its business category shortly after release in 2016. It is a perfect companion book *to Bad Company/Good Company*, offering simple additional tactics to help promote safety as the first step in the journey to a culture of winning.

https://www.amazon.com/Leading-People-Safely-Business-Battlefield/dp/1635051363/ref=asap_bc?ie=UTF8